The young Einstein was haunted by a strange thought: would his image appear in a mirror whilst he and the mirror were moving at the speed of light? His questions were resolved with the creation of the theory of special relativity. *Einstein's Mirror* is a book on relativity with a difference. Following the successful format of their earlier book, *The Quantum Universe*, the authors blend a simple, non-mathematical account of the underlying theory of special relativity and gravitation with a description of the way these theories have been triumphantly supported by experiment.

This book presents a dramatic account of relativity and its immense value in understanding the universe. Novel features include detailed accounts of the role relativity has played in atomic and nuclear physics – from John Dalton's atomic model to Los Alamos and the atomic bomb. The book describes the varied applications of relativity, from satellite navigation systems, particle accelerators and nuclear power to anti-matter, black holes and the origin of the universe. A final chapter looks at the influence of ideas from Einstein's relativity on the development of science fiction from H. G. Wells' *The Time Machine* to *Star Wars* and *Star Trek*. The text is enlivened and supported by a superb collection of colour and black-and-white photographs and by a sprinkling of amusing anecdotes.

Einstein's Mirror provides a fascinating and accessible introduction to one of the most exciting and important scientific discoveries of the twentieth century. Final-year students at school, general readers with an interest in science and undergraduates in science subjects will all be able to enjoy and benefit from this novel exposition.

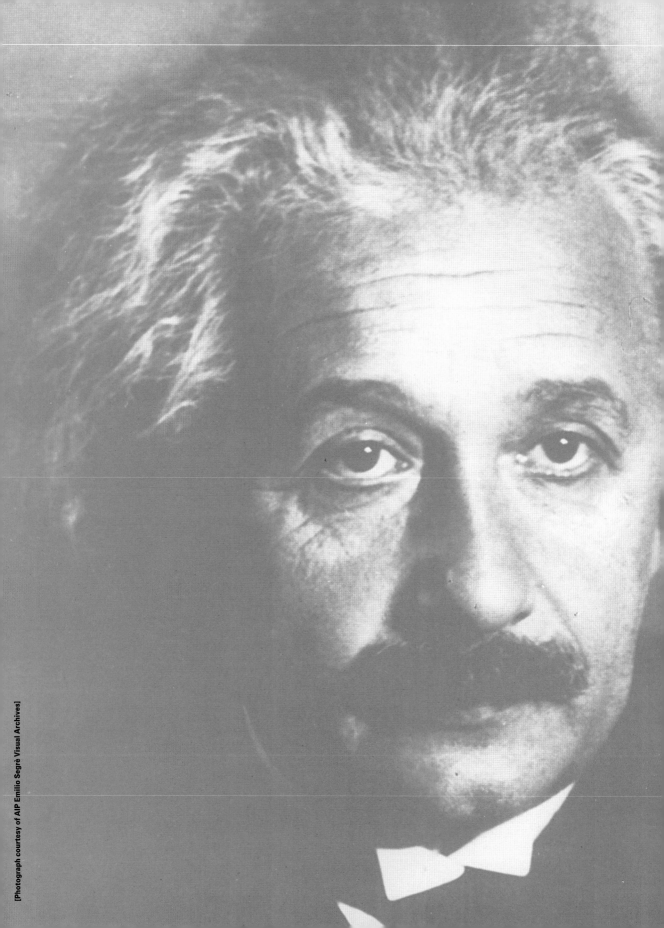

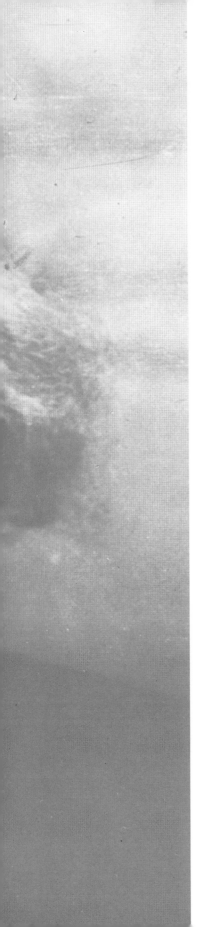

Einstein's Mirror

Tony Hey, Professor
Department of Electronics and Computer Science,
University of Southampton

Patrick Walters, Senior Lecturer
Department of Adult Continuing Education,
University of Wales, Swansea

PUBLISHED BY THE PRESS SYNDICATE OF THE UNIVERSITY OF CAMBRIDGE
The Pitt Building, Trumpington Street, Cambridge CB2 1RP, United Kingdom

CAMBRIDGE UNIVERSITY PRESS
The Edinburgh Building, Cambridge CB2 2RU, United Kingdom
40 West 20th Street, New York, NY 10011-4211, USA
10 Stamford Road, Oakleigh, Melbourne 3166, Australia

First published in 1997

Printed in the United Kingdom at the University Press, Cambridge

Typeset in Monotype Plantin light 9½/12 and Univers

A catalogue record for this book is available from the British Library

Library of Congress Cataloguing in Publication data

Hey, Anthony J. G.
  Einstein's mirror / Tony Hey and Patrick Walters.
    p. cm.
  Includes bibliographical references and index.
  ISBN 0-521-43504-8. – ISBN 0-521-43532-3 (pbk.)
    1. Relativity (Physics)   2. Gravitation.   3. Cosmology.
I. Walters, Patrick, 1949– . II. Title.
QC173.55.H49 1997
530. 11–dc21 96–51097 CIP

ISBN 0 521 43504 8 hardback
ISBN 0 521 43532 3 paperback

Contents

Preface

Encouraged by the success of *The Quantum Universe*, we have tried to adopt a similarly pragmatic approach in this sister volume on Albert Einstein's relativity. Our goal is not only to present the essential ideas of both special and general relativity as simply as possible, but also to demonstrate how the predictions of these theories are verified by the results of experiments. Special relativity is concerned with uniform motion, and does away with Isaac Newton's notion of 'absolute time': it makes startling predictions for objects and observers moving at very high speeds. General relativity, on the other hand, is concerned with accelerations: it turns out to be a theory of gravity which has had a profound impact on our modern view of the universe.

Since our aim is to introduce as many people as possible to the strange world of relativity we have deliberately used a minimal amount of mathematics in the text. Some simple derivations requiring no more than high school maths have been relegated to an appendix for the curious. Needless to say, any book about both special and general relativity must also be to some extent about the physicist who almost single-handedly created these theories. Einstein's legacy is truly remarkable – both inside and outside of physics – and we hope to have captured some flavour of the man through the quotations and stories that accompany the text.

The predictions of quantum mechanics – the laws of physics apparent at very small dimensions – can sometimes be very strange when compared with our everyday experiences. Nonetheless, although quantum mechanics provides an unfamiliar description of the invisible, sub-atomic world, it also makes possible many features of modern life that we now take for granted. Transistors and silicon chips, lasers and superconductors, electron microscopes and particle accelerators, all testify to the phenomenal success of quantum mechanics in providing us with a practical understanding of electrons, atoms and nuclei. It is much more difficult to point to such manifest successes of relativity. Special relativity only predicts significant deviations from Newton's laws when speeds comparable with the speed of light are involved. Since light travels at some 300 million metres/second, any effects due to Einstein's special relativity are negligible for objects moving at the speeds we encounter every day. Nevertheless, it is possible to perform experiments to verify Einstein's prediction for the relativistic slowing down of time – known as 'time dilation' – and the similarly striking prediction for the variation of an object's mass with its velocity. Indeed, both these effects are incorporated as essential elements of the design and daily operation of modern elementary particle accelerators. Besides describing such classic tests of special relativity, a novel feature of this book is a detailed account of its implications for both atomic and nuclear physics. Paul Dirac's prediction of anti-matter, and the Manhattan Project to develop the atomic bomb, were both made possible by the insights into mass and energy given to us by Einstein.

Einstein's three original tests of general relativity have now been subjected to detailed scrutiny for more than half a century. During this time, the technology available to experimental physicists and astronomers has changed dramatically. In addition, Irwin Shapiro has proposed a fourth test for general relativity. Despite some worries, Einstein's gravitational theory has come through all the tests with flying colours and has beaten off all challenges. General relativity is now beginning to be used by astrophysicists as a trusted tool to assist them in their exploration of the universe.

Many analyses have attempted to capture the essence of the 'scientific method'. It is our belief that the way in which physics makes progress can best be explained by means of specific examples. In the history of relativity there are several interesting 'case studies' that resulted in the eventual abandonment of a supposedly scientific theory. The story of the 'luminiferous aether', and Einstein's realization that physics was much simpler without it, is central to the theory of special relativity. Similarly, the decline of the 'phlogiston' theory of fire and of the 'caloric' theory of heat provides us with object lessons as to what constitutes a truly predictive theory. The battle between supporters of thermodynamics and of atoms – with the final victory belonging to the 'atomicists' – shows that progress is certainly not made in the same way as textbook presentations. Moreover, as recounted in some of the anecdotes in this book, a combination of luck and 'physical intuition' has often played a seemingly vital role in major physics discoveries. What makes science work is the possibility of proving a theory wrong: a theory that cannot be tested is ultimately empty. The currently fashionable Theory of Everything based on superstrings seems to come dangerously close to being untestable. It has been suggested that young people dislike following science at school and university because of the absence of debate – there is instead just a long and well-mapped path to the frontiers of research. If this is true, it is a great pity: the whole essence of science lies in proving a theory wrong or in proposing a new theory to be shot down.

The approach to physics in general, and to relativity in particular, that we believe best characterizes the scientific method has been summed up memorably by the late Richard Feynman:

> In general we look for a new law by the following process. First we guess… No! Don't laugh – it's really true. Then we compute the consequences of the guess to see if this law we guessed is right – what it would imply. Then we compare those computation results to nature - or, we say, to experiment, or experience – we compare it directly with observation to see if it works. If it disagrees with experiment, it's wrong. In that simple statement is the key to science. It doesn't make any difference how beautiful the guess is, it doesn't make any difference how smart you are – who made the guess, or what his name is. If it disagrees with experiment, it's wrong. That's all there is to it.

Johannes Kepler's struggles with the inconvenient accuracy of Tycho Brahe's astronomical measurements epitomize this process. After working through an incredible maze of theoretical meanderings, Kepler was forced to recognize that Aristotle's perfect circles were definitely wrong and that the planets preferred orbits that were ugly ellipses. Such an insight, which was able to clear away centuries of folklore and prejudice, usually appears obvious in retrospect. Kepler summed up his tortuous path to discovery with the words: 'Ah, what a foolish bird I have been!'

In similar fashion, Einstein's contributions to physics overturned centuries of unconscious prejudice about the fabric of space and time. In a thoughtful essay entitled 'How can we be sure that Albert Einstein was not a crank?', the writer and physicist Jeremy Bernstein considers the problem of how he would have reacted to receiving Einstein's four 1905 papers in the mail. Who was this twenty-six-year-old patent examiner – technical expert 'third class' – without a physics doctorate, who dared challenge the very foundations of the physics? Why were his papers not immediately consigned to the bin? In answering these questions, Bernstein identifies two criteria to distinguish new physics from crank theories. To the first, Feynman's 'predictiveness', he adds the criterion of 'correspondence'. By 'correspondence', Bernstein means that any radical new theory must also have an explanation of why the 'precedent' science – the science that has gone before – appeared to work. For example, if Newton was so wrong about absolute space and time, how was it that Newtonian mechanics and Newtonian gravity have served physics so well? In Einstein's theories, Newtonian physics emerges as a well-defined approximation, valid for most everyday circumstances. Relativity thus passes Bernstein's correspondence test. As we demonstrate in this book, relativity also passes the second criterion of predictiveness: both theories made specific predictions that were subsequently verified by experiment. According to Feynman's yardstick, relativity is alive and well.

Unsurprisingly perhaps, this project has taken much longer and has involved more hard labour than we anticipated. A large debt of thanks is owed to our families, Marie and Jessie, Nancy, Jonathan and Christopher, for their supreme tolerance during the last few years. We hope they will think their sacrifices worthwhile. Thanks are also due to many colleagues at Southampton and Swansea but especially to Garry McEwen for his tireless reading of incomplete manuscripts and for his many suggestions for improvements in presentation. Finally, we thank Rufus Neal and the staff at Cambridge University Press for their patience and help in putting this book together; Irene Pizzie for her vigilant copy-editing; Harriet Millward for tireless picture research; Phil Treble for his excellent design and layout; and Marie Walters for her assistance with the artwork.

1 A revolution in time

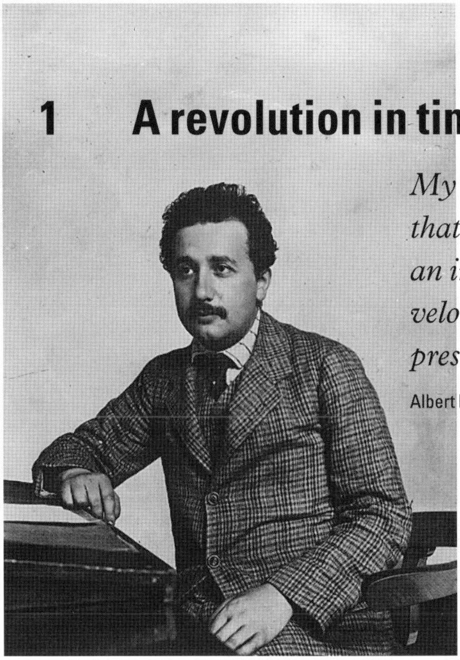

My solution was really for the very concept of time, that is, that time is not absolutely defined but there is an inseparable connection between time and the signal velocity. Five weeks after my recognition of this, the present theory of special relativity was completed.

Albert Einstein, Kyoto Address, 1922

Figure 1.1 Einstein sitting at his desk in the Berne patent office, where he worked from June 1902 until October 1909. He obtained the job on the strength of a recommendation to the director from the father of Marcel Grossmann, his friend and fellow student at the Federal Institute of Technology in Zurich. Einstein was unable to obtain a university position after his degree because he had antagonized one of his professors, Heinrich Weber. As a result, after completing his degree, Einstein had a miserable time being either unemployed or engaged on temporary teaching jobs until he obtained the position at the patent office. [Permission granted by the Albert Einstein Archives, The Jewish National and University Library, The Hebrew University of Jerusalem, Israel.]

Einstein's revolution

The famous Russian scientist Lev Landau used to keep a list of names, in which he graded physicists into 'leagues'. The first division contained the names of physicists such as Niels Bohr, Werner Heisenberg and Erwin Schroedinger, the founding fathers of modern quantum physics, as well as historical 'giants' such as Isaac Newton. He was rather modest about his own classification, grading himself 2½, although he later promoted himself to a 2. Most working physicists would be happy even to make it into Landau's fourth division: David Mermin, a well-known and perceptive American physicist, once wrote an article entitled 'My life with Landau: homage of a 4½ to a 2'. What is the point of this story? The point is that any book about relativity is inevitably also about Albert Einstein, and Einstein was a remarkable physicist by any standard. Landau, in fact, created a special 'superleague' containing only one physicist, Einstein, whom he classified uniquely as a ½. Thus, the popular opinion that Einstein was the greatest physicist since Newton is widely shared among professional physicists.

When Einstein wrote about 'The wonderful events which the great Newton experienced in his younger days…', and commented that, to Newton, Nature was 'an open book', he could well have been writing about himself. Newton's miraculous years of discovery occurred during the Great Plague in 1665–66, when he retired to work in isolation at his home in the Lincolnshire village of Woolsthorpe. For Einstein, the miraculous years of daring innovation were spent in Switzerland, quietly working at the Patent

Figure 1.2 The circle of friends called the 'Olympia Academy' of Berne, 1902. From left to right: Conrad Habicht, Maurice Solovine and Albert Einstein. Together they discussed philosophy, physics and literature. In later life, Einstein wrote to Solovine of those days 'when we ran our happy "academy" which after all was less childish than those respectable ones which I got to know later from close in.' [Courtesy Bundesamt für Kultur/Schweizerische Landesbibliothek.]

Office in Berne. Without any contact with the professional physics community, Einstein, in 1905, published not only the papers that spelled out the special theory of relativity, but also papers that laid the foundation for quantum theory. It was for his work on quantum theory that he won the Nobel Prize for Physics some years later – not for relativity. It is perhaps not surprising that Einstein, after 1905, should wonder whether his creative days were over. He wrote to his close friend Maurice Solovine: 'Soon I will arrive at the stationary and sterile age where one laments over the revolutionary mentality of the young.' Actually, Einstein was still only in his mid-twenties, and his supreme achievement – the creation of the theory of general relativity – was yet to come. Although Einstein obtained his first intuitive glimpse of a new theory of gravity in 1907, it took him ten more years of hard work to transform this into the theory of general relativity that we know today. In the following chapters, we shall attempt to explain what physicists understand by both 'special' and 'general' relativity, and to give some idea of the revolution each brought to the way in which physicists view the world.

It was not until after the First World War that Einstein, then a professor in Berlin, suddenly acquired world-wide fame. His prediction of the bending of light passing near the Sun was confirmed by a British expedition sent to photograph the total eclipse of the Sun in 1919. Headlines in *The Times* proclaimed 'Revolution in Science – New Theory of the Universe – Newtonian Ideas Overthrown'. Before we have a look at what Einstein said about the world, we should perhaps remember what turbulent times these were in Europe. The Russian Revolution was still sending shock waves around the world, and even the art world was experiencing an unparalleled upheaval with the birth of the Dada movement. Curiously, both these revolutions had connections with Switzerland, where Einstein had completed his physics education. Was the parallel birth in Berne of relativity and the revolutions in politics and art just a coincidence? Zurich must have been a stimulating and exciting place for a young student in those days – certainly a long way from the 'safe' image we have of Switzerland today.

The sociologist Lewis Feuer has indeed suggested that Einstein's theories of relativity were nurtured by the revolutionary culture of Zurich. The name *relativity* is perhaps a clue, since, in many ways, it is strangely

Zurich – hot-bed of revolution

Lenin and many other revolutionaries had found safe havens in Switzerland because of its policy to grant political asylum. In the eyes of many of the leaders of other European countries, this made peaceful, democratic Switzerland a very dangerous place. Indeed, in 1873, the Czarist Russian government had demanded that students 'immediately abandon the terrible city of Zurich'. And perhaps they were right to be worried – it was from Zurich, in the spring of 1917, that Lenin and a band of his associates were allowed, with the connivance of the German authorities, to leave for Russia in a sealed train.

The Dada art movement began in Zurich in 1916 as a revolt against the traditional values that were embraced by a society that could countenance the mass slaughter which took place during the First World War. Dada was deliberately provocative and outrageous and it sought to shock people out of a state of complacency. The name, meaning 'a child's wooden hobbyhorse,' is thought to have been chosen at random from a German–French dictionary. The movement was launched at a nightclub called the Cabaret Voltaire by the playwright Hugo Ball and the singer Emmy Henning. The public outcry was such that the 'cabarets' were soon stopped, but Dada, like Marxist-Leninism, continued to flourish and spread across the world. Eventually, some of the Dada artists joined with others to form the Surrealist art movement, of whom Salvador Dali was a prominent member.

Figure 1.3 The laboratory at the Zurich Federal Institute of Technology where Einstein was an undergraduate. [Permission granted by ETH – Bibliothek.]

inappropriate. All too often Einstein's theory is over-simplified to mean that 'everything is relative'. Einstein himself is guilty of using this aspect of relativity theory when writing in *The Times*:

> By an application of the theory of relativity to the taste of readers, today in Germany I am called a German man of science and in England I am represented as a Swiss Jew. If I come to be regarded as a bete noire, the descriptions will be reversed and I shall become a Swiss Jew for the Germans and a German man of science for the English.

The Times retorted:

> We concede him his little jest; but note that, in accordance with the general tenor of his theory, Dr. Einstein does not supply an absolute description of himself.

The theory of relativity has no connection with either the social or moral relativity as expressed by the revolutionary movements in politics or art. Indeed, since relativity theory makes very specific and definite predictions that agree with experimental observations, 'absolute theory' would be an equally valid name. It is intriguing to speculate that Einstein, who had great empathy with the oppressed of the world, may have unconsciously had the idea of social relativity in his mind when naming his startling new vision of the physical world. In this sense, it might be said that the theory of relativity born in the quiet Berne patent office is the most abiding achievement of the revolutionary movements that emerged from Switzerland at the turn of the century. The story of how this revolution challenged, and continues to challenge, our preconceived ideas about space and time is the theme of this book.

Einstein wrote in his autobiography:

> …sometimes we 'wonder' quite spontaneously about some experience. This 'wondering' seems to occur when an experience comes into conflict with a world of concepts that is already fixed in us.

Figure 1.4 A view of Zurich, Switzerland, where Einstein studied at the Federal Institute of Technology. He became a Swiss citizen and even volunteered for military service but was rejected as unfit. The commercial city of Zurich played a surprising role in the revolutionary movements in Europe at the beginning of the twentieth century. [Courtesy Swiss Tourism.]

Relativity causes much 'wonder' in the struggle to come to terms with its ideas. The problem lies in trying to reconcile Einstein's view of time and space with our own intuition drawn from our experiences in everyday life. Relativistic ideas are so alien to us that they seem unbelievable. Like quantum mechanics, relativity makes predictions seemingly entirely at odds with our own experiences. However, as in the case of quantum theory, scientific experiment validates the predictions of Einstein's relativity theory. We have no option, therefore, but to modify and enlarge our thinking to include these ideas. To convince the reader that the strange predictions of relativity have been carefully checked and are verified daily in many ways in the world around us is the central purpose of this book.

THE GLORIOUS DEAD.

KING'S CALL TO HIS PEOPLE.

ARMISTICE DAY OBSERVANCE.

TWO MINUTES' PAUSE FROM WORK.

The King invites all his people to join him in a special celebration of the anniversary of the cessation of war, as set forth in the following message :—

TO ALL MY PEOPLE.

BUCKINGHAM PALACE.

Tuesday next, November 11, is the first anniversary of the Armistice, which stayed the world-wide carnage of the four preceding years and marked the victory of Right and Freedom. I believe that my people in every part of the Empire fervently wish to perpetuate the memory of that Great Deliverance, and of those who laid down their lives to achieve it.

To afford an opportunity for the universal expression of this feeling it is my desire and hope that at the hour when the Armistice came into force, the eleventh hour of the eleventh day of the eleventh month, there may be, for the brief space of two minutes, a complete suspension of all our normal activities. During that time, except in the rare cases where this may be impracticable, all work, all sound, and all locomotion should cease, so that, in perfect stillness, the thoughts of every one may be concentrated on reverent remembrance of the Glorious Dead.

No elaborate organization appears to be necessary. At a given signal, which can easily be arranged to suit the circumstances of each locality, I believe that we shall all gladly interrupt our business and pleasure, whatever it may be, and unite in this simple service of Silence and Remembrance.

GEORGE R.I.

YUDENITCH'S LOSSES.

RETIRING IN GOOD ORDER.

(FROM OUR SPECIAL CORRESPONDENT.)

HELSINGFORS, Nov. 6.

A Bolshevist detachment has crossed the river Plyusa in the direction of Gdoff [Yudenitch's southern base].

General Yudenitch's retirement has been effected in good order. His total losses in men are computed at 4,000—a total more than made good by enrolments from prisoners. The loss of material was small, and all the Tanks were brought back.

The Foreign Legion, which numbered about 200, lost three-fourths of its strength, having taken part in the severe fighting at Yamburg at the beginning of the offensive. This Legion was formed at Archangel, and was chiefly composed of Russians. It earned its name because it was officered originally by the French. Since its transfer to the North-West Front it has contained also Danes, Swedes, and Swiss.

Unfortunately, the mistakes made in the previous offensive were repeated in Yudenitch's last advance. There was not sufficient unity in the higher command, nor sufficient care taken to consult the feelings of the population. Further, a complete picked division at the disposal of General Laidoner (the Esthonian Commander-in-Chief) for use against Krasnaya Gorka could not be sent, presumably because of the unsympathetic attitude of Russian military circles towards the Esthonians.

November 9 has been named as the date for the resumption of the discussions between the Baltic States on the question of peace negotiations with the Bolshevists. It is reported that an envoy has been dispatched by the Polish Government to sound the Baltic States on a plan of common action, not excluding the possibility of Poland's making peace with the Bolshevists.

RED ADVANCE.

The Bolshevist report of November 5 shows that on Yudenitch's left flank and in the centre the Red Army has made little progress since Sunday. The Reds claim to be no more than 10 miles west of Gatchina.

On Yudenitch's right flank, however, the Red advance is marked. The Bolshevists have now almost complete possession of the railway from Gatchina to Pskoff and from the east and the south they are marching towards Gdoff, Yudenitch's base on Lake Peipus. "We are," says the *communiqué*, "uninterruptedly advancing."

FINLAND DECLINES TO INTERVENE.

ALLIES' ATTITUDE.

(FROM OUR CORRESPONDENT.)

ABO, Nov. 5.

The Finnish Government has replied to M. Lianosoff, the head of the North-West Russian Government, stating that it is unable to grant his request for Finnish help against the Bolshevists.

Discussing the question of Finnish intervention the Helsingfors newspaper *Hufvudstadsblad* writes that the Government's decision was scarcely unexpected.

The question (it says) of Finland's participation in the struggle against Bolshevism must be dependent upon due security being given us that the materials necessary for the enterprise are placed at our disposition that the cost of the military expedition should be repaid us, and also that should large contingents of Bolshevist troops be sent against our troops they should not be left to their own resources but could depend upon the effective help of the Powers. From the Government's answer it is clear that at present these guarantees do not exist.

No promises from the Entente States have been given or any request made to Finland to give active support to the White Russians. A telegram received by the Government from Paris categorically explains that France is not willing to give assistance and the French Government does not wish to exhort Finland to action as France is not willing to invest itself with the responsibility.

*** We understand that this explanation of the attitude of France also faithfully represents that of the British Government.

KOLTCHAK LEAVING OMSK.

STEADY BOLSHEVIST ADVANCE.

OMSK, Oct. 30 (Delayed).—The Civil Government is evacuating Omsk. Admiral Koltchak's army is retreating on the whole front.—*Reuter*

For the past fortnight the Eastern Bolshevist Army, reinforced by troops from Turkestan and Central Russia, has been conducting an energetic offensive against the Siberian Army. The right wing of the Reds, which had been driven back almost to Kurgan, made a determined attack, recaptured Petropavlovsk—where the Ishim

river crosses the Siberian Railway—and by the last report was 27 miles east of that town—that is, some 170 miles west of Omsk. In the centre the Reds are also advancing towards Omsk; on their left (north) flank, since the capture of Tobolsk, their progress has been slow.

BOLSHEVIST report, Nov. 5 :—

Along the Siberian Railway the Red troops are advancing on Bezrukobsky, 14 miles west of Ishim. We have captured 1,200 prisoners, and trophies which have not yet been counted.

On the night of November 5 the Red troops triumphantly occupied the town of Ishim.—*Wireless Press.*

REVOLUTION IN SCIENCE.

NEW THEORY OF THE UNIVERSE.

NEWTONIAN IDEAS OVERTHROWN.

Yesterday afternoon in the rooms of the Royal Society, at a joint session of the Royal and Astronomical Societies, the results obtained by British observers of the total solar eclipse of May 29 were discussed.

The greatest possible interest had been aroused in scientific circles by the hope that rival theories of a fundamental physical problem would be put to the test, and there was a very large attendance of astronomers and physicists. It was generally accepted that the observations were decisive in the verifying of the prediction of the famous physicist, Einstein, stated by the President of the Royal Society as being the most remarkable scientific event since the discovery of the predicted existence of the planet Neptune. But there was difference of opinion as to whether science had to face merely a new and unexplained fact, or to reckon with a theory that would completely revolutionize the accepted fundamentals of physics.

SIR FRANK DYSON, the Astronomer Royal, described the work of the expeditions sent respectively to Sobral in North Brazil and the island of Principe, off the West Coast of Africa. At each of these places, if the weather were propitious on the day of the eclipse, it would be possible to take during totality a set of photographs of the obscured sun and of a number of bright stars which happened to be in its immediate vicinity. The desired object was to ascertain whether the light from these stars, as it passed the sun, came as directly towards us as if the sun were not there, or if there was a deflection due to its presence, and if the latter proved to be the case, what the amount of the deflection was. If deflection did occur, the stars would appear on the photographic plates at a measurable distance from their theoretical positions. He explained in detail the apparatus that had been employed, the corrections that had to be made for various disturbing factors, and the methods by which comparison between the theoretical and the observed positions had been made. He convinced the meeting that the results were definite and conclusive. Deflection did take place, and the measurements showed that the extent of the deflection was in close accord with the theoretical degree predicted by Einstein, as opposed to half that degree, the amount that would follow from the principles of Newton. It is interesting to recall that Sir Oliver Lodge, speaking at the Royal Institution last February, had also ventured on a prediction. He doubted if deflection would be observed, but was confident that if it did take place, it would follow the law of Newton and not that of Einstein.

DR. CROMMELIN and PROFESSOR EDDINGTON, two of the actual observers, followed the Astronomer-Royal, and gave interesting accounts of their work, in every way confirming the general conclusions that had been enunciated.

"MOMENTOUS PRONOUNCEMENT."

So far the matter was clear, but when the discussion began, it was plain that the scientific interest centred more in the theoretical bearings of the results than in the results themselves. Even the President of the Royal Society, in stating that they had just listened to "one of the most momentous, if not the most momentous, pronouncements of human thought," had to confess that no one had yet succeeded in stating in clear language what the theory of Einstein really was. It was accepted, however, that Einstein, on the basis of his theory, had made three predictions. The first, as to the motion of the planet Mercury, had been verified. The second, as to the existence and the degree of deflection of light as it passed the sphere of influence of the sun, had now been verified. As to the third, which depended on spectroscopic observations there was still uncertainty. But he was confident that the Einstein theory must now be reckoned with, and that our conceptions of the fabric of the universe must be fundamentally altered.

At this stage Sir Oliver Lodge, whose contribution to the discussion had been eagerly expected, left the meeting.

Subsequent speakers joined in congratulating the observers, and agreed in accepting their results. More than one, however, including Professor Newall, of Cambridge, hesitated as to the full extent of the inferences that had been drawn and suggested that the phenomena might be due to an unknown solar atmosphere further in its extent than had been supposed and with unknown properties. No speaker succeeded in giving a clear non-mathematical statement of the theoretical question.

SPACE "WARPED."

Put in the most general way it may be described as follows: the Newtonian principles assume that space is invariable, that, for instance, the three angles of a triangle always equal, and must equal, two right angles. But these principles really rest on the observation that the angles of a triangle do equal two right angles, and that a circle is really circular. But there are certain physical facts that seem to throw doubt on the universality of these observations, and suggest that space may acquire a twist or warp in certain circumstances, as, for instance, under the influence of gravitation, a dislocation in itself slight and applying to the instruments of measurement as well as to the things measured. The Einstein doctrine is that the qualities of space, hitherto believed absolute, are relative to their circumstances. He drew the inference from his theory that in certain cases actual measurement of light would show the effects of the warping in a degree that could be predicted and calculated. His predictions in two of three cases have now been verified, but the question remains open as to whether the verifications prove the theory from which the predictions were deduced.

Figure 1.5 Articles from the *Times* of November 7th, 1919. They include a report on Eddington's eclipse expedition that confirmed Einstein's prediction of the bending of light rays by the gravitational influence of the Sun. [*The Times*, © Times Newspapers Limited, 1919. Provided by the Newspaper Library of the British Library.]

How did Einstein discover relativity? The crucial insight that led Einstein to the theory of relativity only arrived after much hard work and after he had attained a deep understanding of both the successes and limitations of the 'classical' theories of Newton and Maxwell. Curiously, Einstein explained his discovery in terms of his 'slow' development:

> The normal adult never bothers his head about space-time problems. Everything there is to be thought about it, in his opinion, has already been done in early childhood. I, on the contrary, developed so slowly that I only began to wonder about space and time when I was already grown up. In consequence I probed deeper into the problem than an ordinary child would have done.

In particular, at the age of sixteen, Einstein puzzled over the consequences of travelling at the speed of light. Nothing in nineteenth century science limited the maximum speed attainable, so this fantasy seemed possible, if a little impractical. However, just thinking about the implications of travel at light speed posed several intriguing problems. One of these we have called 'Einstein's Mirror': what would you see if you and the mirror you were

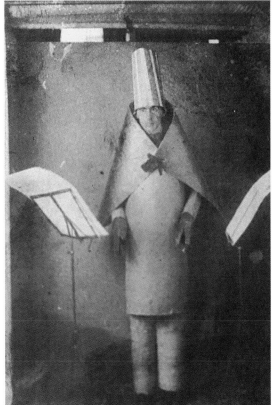

Figure 1.6 Hugo Ball (1886–1927) was a German pacifist and one of the founders of the Dada art movement. In many ways, Dada was a reaction against the horrors and madness of the First World War. Dada celebrates chance – Tristan Tzara, another of its founders, declared in a manifesto that 'Dada means nothing'. In fact, Dada represented a riotous rebellion against all the old values, artistic conventions and political certainties of the past. Ball married the singer Emmy Henning, and later in life renounced all excess and retired to live a devout and simple life among the Swiss peasants. His autobiography was called *Flight from Time*.

▲
Figure 1.7 Friedrick Adler (1879–1960), physicist and Marxist revolutionary. Einstein lived for a while in the same house as Adler while they were fellow physics students and friends. In 1909, Adler selflessly made way for Einstein to obtain his first academic post at Zurich University. This is a curious parallel of Isaac Barrow's resignation of his Professorship at Cambridge in 1669 in favour of Isaac Newton. In 1916, Adler shot the Austrian Prime Minister. His defence at his trial was a strange mixture of relativity in morals and physics combined with dogmatic Marxism. Einstein offered to write a letter in his defence.

looking into were both moving at the speed of light? This is an example of a 'thought' experiment – usually known in physics as a 'gedanken' experiment, from the German for thought. Einstein used such imaginary experiments to great effect all his life, including in his famous (but ultimately unsuccessful) attempts, in an exchange of letters with Niels Bohr, to prove that quantum mechanics was inconsistent. An account of Einstein's role in the foundation of quantum mechanics is contained in the companion volume to this book, *The Quantum Universe*.

Einstein solved his mirror problem with his creation of 'special relativity'. With the example of the mirror moving at the speed of light Einstein had put his finger on the key to the whole problem. Even though Einstein was not the only scientist to be concerned with the problem of light and how its velocity might be affected by the motion of the observer, it was Einstein, and Einstein alone, who realized that the solution to the problem involved a radical re-thinking of the concept of time. The first seven chapters of this book will be devoted to exploring the consequences of this new vision of space and time. This is the content of Einstein's 'special relativity' theory, and we will show how it finds routine application in physics, chemistry and medicine. In particular, the huge particle accelerators in

Figure 1.8 Michele Angelo Besso (1873–1955) and his wife Ann, the daughter of Einstein's old school master, Jost Winteler, in 1898. Besso was Einstein's closest confidant all his life and a fellow patent clerk at Berne. It was to Besso that relativity theory was first explained. Einstein said of Besso: 'I could not have got a better sounding board in the whole of Europe.' Recalling their time together at the patent office, Einstein wrote to Besso of: '... that cloister where I hatched my most beautiful ideas and where we had such good times together.'

Europe, the USA and Russia depend on special relativity for their very operation and design.

The situation was rather different for the theory of 'general relativity': essentially a theory of gravity. In a sense, one can say that Einstein's special theory modified Newton's laws of motion and that his general theory modified Newton's theory of gravity. Einstein described the inspiration which led to his theory of general relativity as 'the happiest thought of my life'. He explained his happy thought as follows:

> I was sitting in a chair in the patent office at Berne when all of a sudden a thought occurred to me: 'If a person falls freely he will not feel his own weight.' I was startled. This simple thought made a great impression on me. It impelled me towards a theory of gravitation.

This image of a 'falling man' preoccupied Einstein for many years. After eight years of intense effort and imagination, he was able, in 1915, to convert this image into the fully-fledged theory of general relativity. The new theory quick-

Figure 1.9 Einstein (at the right of the front row) in the classroom at Aarau, Switzerland. His teacher is Dr Jost Winteler, whose daughter married his friend Besso. Einstein had disliked the rigid disciplined education he received in Munich. In the free democratic atmosphere of Switzerland, he acquired the self-confidence that remained with him all his life. It was while at school in Aarau that Einstein first conceived of the problem of pursuing a light beam. From Fluckiger, *Einstein in Bern*, 1974. [Courtesy AIP Emilio Segrè Visual Archives.]

ly found observational success with the British eclipse expedition mentioned earlier. Although it was this success that caught the imagination of the popular press and catapulted Einstein onto the world stage, unlike special relativity, general relativity found very few other areas of application: it became rather like a 'temple' of modern physics, revered and respected, but seldom visited. All this changed dramatically in the 1960s. Developments in atomic clocks, space travel and radio astronomy breathed new life into the subject, and new and more accurate tests of Einstein's ideas were devised and undertaken. Quasars and pulsars were discovered, and a new generation of physicists began to explore seriously the implications of the strange 'black holes' predicted by the general theory. The discovery of the microwave background radiation in 1965 and its interpretation as a relic of the 'Big Bang' creation of the universe, led to renewed interest in the use of Einstein's general theory of relativity as a tool for modern cosmology. Experiments are in progress at the time of writing that attempt to detect the gravitational waves predicted by Einstein. It is because of all these new 'post-Einstein' developments that his general theory is now firmly in the vanguard of modern experimental and theoretical research. These topics form the basis for chapters 8–10.

No book about relativity would be complete without at least a passing look at the problem that occupied a large part of the remainder of Einstein's research life. This was his dream of a 'unified theory' of all natural laws. During Einstein's lifetime, this research was not in the mainstream, and was very distant from the exciting applications of quantum mechanics and nuclear physics. Today, however, such a search is now an active research area for theoretical physicists. Einstein made the geometry of space-time the key to gravity in general relativity. The extension of similar geometrical ideas to other areas of physics has been one of the important consequences of Einstein's efforts. Chapter 10 describes the state of these attempts as they exist in the late 1990s.

Time and clocks

It might appear possible to overcome all the difficulties attending the definition of 'time' by substituting 'the position of the small hand of my watch' for 'time'. And in fact such a definition is satisfactory when we are concerned with defining time exclusively for the place where the watch is located; but it is no longer satisfactory when we have to...evaluate the times of events occurring at places remote from the watch.
Albert Einstein, 'The electrodynamics of moving bodies', 1905

Many people have had the experience of returning from some routine chore with no conscious recollection of having completed it. This is common when driving, for example, when part of the mind seems to take over driving the car while the conscious mind is occupied with other things: control is only reasserted if something unusual happens. Miraculously, the task has been accomplished on 'automatic pilot', without conscious attention. One strange feature of this experience is the absence of any sense of the passage of time. This 'time gap' phenomenon makes apparent what is significant for our sense of time: conscious attention to a series of physical or mental changes which act as time markers. Is 'time' purely reducible to an ordered sequence of events such as the passage of day and night, or to the biological processes of ageing? Or is 'time' some more fundamental inner property of Nature?

If we turn over an hour-glass, the interval of time taken for the sand to run out is assumed to be the same, no matter how often we repeat the process. (In practice, although we may try to duplicate the same event, things can go wrong – such as the sand becoming damp and sticking.) The essence of a clock is that it is a device that repeats the same action and

Figure 1.10 A reconstruction of the astronomical clock made by Giovanni de Dondi of Padua in 1350. This beautiful mechanical device took Dondi sixteen years to complete. On each of the seven faces, the path of a 'planet' (as seen from the Earth) was depicted. In those days, the Sun and the Moon were regarded as planets and the Earth was imagined to be fixed at the centre of the universe. When this clock was made, mechanical clocks had only recently been invented. Notice the close association between astronomical and mechanical time represented in this early clock. [Courtesy National Museum of American History, Smithsonian Institution.]

Figure 1.11 Pisa cathedral with its ornate chandelier. In order to light the candles, a monk draws the chandelier to one side with a long pole. After completing this task, the monk releases the chandelier, which swings in great arcs for a considerable time. One Sunday morning in 1581, Galileo noticed that the time for a complete swing remained the same as the speed of motion and size of the swing decreased. Galileo timed the period of the swing using his pulse. This is said to be the origin of the idea behind the pendulum clock.

counts the number of occurrences of the action. In this way, a clock can measure the passage of time. One simple clock could be a device for counting the number of heartbeats. Needless to say, this would not be a very accurate clock since our heartbeat varies considerably when we are excited or frightened, or take vigorous exercise. Nevertheless, it is clear how we could use many types of everyday events to act as indicators for the passage of time.

The most familiar marker for the passage of time is the rising and setting of the Sun. It would be difficult to use sunrise and sunset as the basis for a clock since the amount of 'day-time', as opposed to 'night-time', varies with the seasons. Even the sum of day- and night-time to define a day – the interval between the times when the Sun is at the highest point in the sky – varies throughout the year. The facts that the orbit of the Earth is not a circle but an ellipse, and that the axis of rotation of the Earth is tilted away from the vertical, mean that the 'solar' day varies by up to fifteen minutes. Perhaps 'star-time', which defines the rotation period of the Earth by the motions of the stars around the sky, would be a better choice for a standard time. Unfortunately, this is also not completely uniform, and is affected, for example, by the tidal influence of the Moon. Two hundred million years ago (when only dinosaurs needed to tell the time), the length of the day was only about 23 hours, and a year would have contained about 380 days. What about using the time taken for the Earth to orbit round the Sun as our standard measure? In fact, even this time undergoes small alterations caused by the subtle gravitational effects of the other planets in the solar system. These corrections to the length of the year can be calculated using Newton's laws of motion and gravity. Implicit in these laws is the concept that Newton called 'absolute time'. Absolute time implies that there is a single universal time that is valid throughout the universe. It is this idea that was overturned by Einstein.

Our commonsense notion of time has gradually become less reliant on the Sun and the stars and other natural phenomena, and has evolved into a more 'abstract' view based on the widespread use of watches and clocks. This development began about 300 years ago with the introduction of accurate mechanical clocks, and has continued to the present day with quartz crystal watches – now cheap enough to be given away in marketing promotions. In the scientific laboratory, the most accurate timing devices are so-called 'atomic clocks'. These use the frequency of the radiation that is generated when an atom makes a quantum jump between two of the very accurately defined atomic energy levels. These are so accurate that, since 1967, the international standard second has been defined with the aid of a caesium-beam atomic clock. The particular energy levels chosen correspond to a frequency of radiation in the microwave region: the international second is defined as 9 192 631 770 oscillations of this radiation. The use of such accurate clocks has revealed tiny changes in the Earth's rotation. To prevent these errors from adding up and causing atomic time to be out of step with our

Figure 1.12 Star trails formed by leaving the camera shutter open for about ten hours. This picture was taken over the dome of the Anglo-Australian telescope, in Siding Spring, New South Wales, Australia, and shows the region near the south pole of the sky. The apparent motion of the stars is caused by the rotation of the Earth. [© Anglo-Australian Observatory; photography by David Malin.]

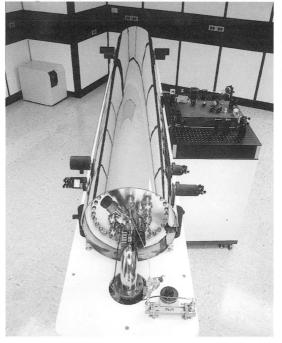

Figure 1.14 NIST-7, an optically pumped atomic clock, at the National Institute of Standards and Technology. Its accuracy is derived from the stability of the vibrational frequencies of the caesium atom. [Courtesy US National Institute of Standards and Technology.]

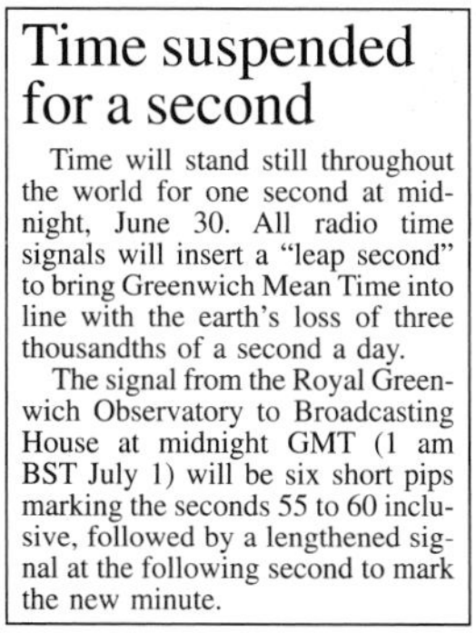

Time suspended for a second

Time will stand still throughout the world for one second at midnight, June 30. All radio time signals will insert a "leap second" to bring Greenwich Mean Time into line with the earth's loss of three thousandths of a second a day.

The signal from the Royal Greenwich Observatory to Broadcasting House at midnight GMT (1 am BST July 1) will be six short pips marking the seconds 55 to 60 inclusive, followed by a lengthened signal at the following second to mark the new minute.

Figure 1.13 A clipping from the *Times* of Wednesday, June 21st, 1972. [*The Times*, © Times Newspapers Limited, 1972. Provided by the Newspaper Library of the British Library.]

calendar year, 'leap seconds' now have to be introduced, usually one per year.

Despite the astonishing precision of atomic clocks, they do not change our basic concept of time in any fundamental way. What Einstein showed was that if we try to compare times of widely separated events, our commonsense view of time must break down. Different observers can measure different times for the same event. Underlying Einstein's breakthrough – in realizing that time was not absolute – is the peculiar nature of the speed of light, which always remains the same, even when measured in different situations. Because these facts about time and light play such a key role in all that follows, it is worth illustrating them as concretely as possible. So let us step forward in time and take a trip on the *Starship Enterprise* – and boldly go to places where the mysteries of time and space may lead to life or death consequences.

MISSION 1: *Moving clocks run slow*

The anti-matter warp drives on the *Starship Enterprise* can propel the spaceship at speeds close to the speed of light, some 300 million metres/second. On mission 1, we travel on the *Enterprise* to a nearby star system at half light speed, and then return to Earth at the same speed. When we return to Earth, we find that thirty years have elapsed, but, according to our ship-board clocks, we believe the total journey time is only twenty-six years. That this seems incredible is only because we have automatically assumed that the clocks on board the *Enterprise* tick at the same rate as those on Earth. This

Figure 1.15 The *Starship Enterprise.* [Copyright © 1988 by Paramount Pictures Corporation.]

example is the essence of the famous 'twin paradox' of relativity. The slowing down of time on the ship applies to all clocks, including the biological mechanisms of ageing. Thus, when you greet your twin brother on your return, you find you are now four years younger than him! A key point in understanding this apparent paradox is to note that the twins do not have identical experiences. Only the twin on board the ship felt the deceleration and acceleration when the ship turned round.

We will consider the implications of such dilemmas in more detail later in the book. For the moment we ask the reader to take on trust the fact that:

moving clocks run slower than stationary clocks

We will look at some of the experimental evidence for this in the last section of this chapter. At speeds much less than the speed of light, the effect of this slowing down of time is extremely small. For example, project Daedalus of the British Interplanetary Society envisages a nuclear-powered spacecraft that could reach about 13% of the speed of light. In the fifty years or so that the ship would take to reach Barnard Star, the on-board clock would have lost about five months compared with its Earth-bound counterparts. For a typical satellite orbiting the Earth, the time retardation due to its velocity is less than one part in 1000 million. However, because of the accuracy of atomic clocks, this time is significant and leads to measurable consequences.

MISSION 2: *Speeds don't add up*

On our next mission aboard the *Starship Enterprise*, we are travelling close to enemy territory and have our photon torpedos at the ready. Suddenly, a red alert sounds: the ship has inadvertently strayed into Klingon-controlled space. Retro-rockets fire and the *Enterprise* makes a rapid retreat at half light speed – but not before the Klingons have launched a deadly photon torpedo towards our ship. Now, commonsense suggests that, since we are retreating at half light speed, and the torpedo is moving at full light speed, the lethal light pulse must be closing in on the ship at half the speed of light. We would be wrong: from the bridge of the *Enterprise*, we would measure the speed of the approaching light torpedo as exactly light speed. This is one of the fundamental cornerstones of relativity: the speed of light is the same *no matter how the observer is moving*. Needless to say, when this surprising behaviour of light was first proposed by Einstein there was considerable resistance to such a radical idea. However, this fundamental property of light has now been fully confirmed by many careful experiments, as we shall see later.

There are more surprises in store. The *Enterprise* decides to return fire with its own salvo of photon torpedos. The Klingon ship which fired on us is now making a hasty retreat, also at half light speed. So, if we are moving in one direction at half the speed of light, and the Klingon ship is moving in the opposite direction also at half the speed of light, our commonsense intuition would suggest that our torpedos travelling at the speed of light could never catch up with the target. Once again, we would be wrong: observers on the Klingon ship see our torpedos approaching their ship at full light speed. Moreover, we see the Klingons moving away from us not at the speed of light but at just 80% of light speed.

We will explore this strange way of adding velocities later in the book, but for the present we ask the reader to note:

> **the speed of light is unaltered by either the motion of the sender or of the observer**

In other words, all observers, regardless of their state of motion, measure the same speed for light. This strange and counter-intuitive behaviour lies at the heart of relativity theory.

One further property of light speed, which no doubt influenced the choice of photon torpedos as a strike weapon, is that the velocity of light is the maximum speed – it cannot be exceeded by any material object. This again is contrary to Newton's laws of motion, where one can seemingly accelerate any matter indefinitely to reach any desired speed. According to Einstein, however, to accelerate a rocket to the speed of light would require an infinite amount of energy. Again, all these strange statements need to be backed up by experiment – and they have been, as we shall see later.

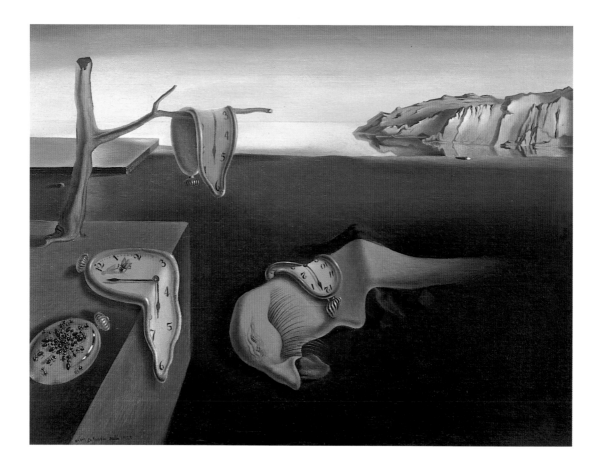

Figure 1.16 *The Persistence of Memory* or *Soft Watches* by Salvador Dali (1931). Oil on canvas, 24.1 × 33 cm. [© 1996 The Museum of Modern Art, New York.]

MISSION 3: *Time is slowed down by gravity*

Having survived its previous encounter with the Klingons, the *Enterprise* continues its exploration of the universe, boldly going where no one has been before. Now, according to general relativity, there exist objects known as 'black holes' – regions of space in which matter has become so compressed and in which gravity is so powerful that not even light rays can escape from them. The *Enterprise* is now approaching one of these objects and must navigate carefully. Anything passing beyond a critical region around the black hole – the so-called 'event horizon' – is doomed never to escape from its gravitational forces. Since no one on board the *Enterprise* wants to undertake such a one-way trip into the unknown, we launch an automatic unmanned probe towards the black hole. At first, we see the probe pick up speed and accelerate towards the black hole, but as it approaches the event horizon, it seems to slow down and appears almost to hover, never quite crossing the critical boundary. Curiously enough, despite this behaviour seen from the *Enterprise*, any unfortunate stowaway aboard the probe would experience only a short time before the probe passed the event horizon and fell into the black hole! We will talk about conditions inside a black hole in a later chapter.

The only point of this adventure that we wish to emphasize here is this:

time is slowed down by gravity

Black holes may seem like very exotic objects remote from everyday experience – and so they are. Not so remote are objects called 'neutron stars' in which matter is not compressed to such a high density. A neutron star still has an extraordinarily high density: a teaspoonful of neutron star matter would weigh about 1000 million tons. Neutron stars are now well-accepted astronomical objects, with a mass comparable with that of our Sun compressed into a volume only 10 miles in diameter. On the surface of a neutron star, clocks would run significantly slower than clocks aboard the *Enterprise*. Unlike the case for black holes, objects launched from the surface of the neutron star with a speed close to that of light would be able to escape from its powerful gravitational attraction.

This slowing-down effect of gravity on time has measurable consequences for satellites in orbit round the Earth. Clocks on the satellite run faster than those on the Earth's surface because the satellite feels less gravitational influence in its orbit high above the Earth. The effect is very small – about one part in 1000 million – and is in the opposite direction to the effect of the satellite's speed on the clock, which slows it down. Modern experiments are now able to verify these predictions of relativity for satellites and rockets.

Experiments with time

We conclude that a balance clock at the equator must go more slowly, by a very small amount, than a precisely similar clock situated at one of the poles under otherwise identical conditions.
Albert Einstein, 'The electrodynamics of moving bodies', 1905

Since Copernicus's time, the observation that the Sun rises in the east has been interpreted as a result of the Earth's rotation. Near the equator, inhabitants of the Earth are rotated in an easterly direction at a speed of about 1000 miles/hour. As recognized by Einstein as long ago as 1905, this motion means that a clock is slowed at the equator relative to an identical clock at the poles. This slowing of time – usually called 'time dilation' in scientific papers – amounts to only around 4 seconds after 1000 years. The advent of atomic clocks with their incredible accuracy have made Einstein's apparently untestable predictions capable of verification by experiment.

We have mentioned the development of atomic clocks earlier in this chapter. By about 1960, it was possible to use them to detect tiny changes in the rotation period of the Earth. Several groups of scientists then realized that the time changes predicted by Einstein could be measured using atomic clocks carried by satellites or rockets. Such experiments were very expensive, and much effort was needed to raise the money. High-level persuasion takes time, and, while this was going on, two physicists, Joseph Hafele and Richard Keating, were able to pre-empt their colleagues and perform a 'bargain-basement' version of the same experiment. They did this by measuring the time dilation effect on atomic clocks carried on board ordinary commercial aircraft flights around the world.

In October 1971, Hafele and Keating flew round the world carrying their atomic clocks as 'hand luggage', first in an easterly direction, then in a westerly direction. Flying both ways around the world and averaging the result removed the effect of the Earth's rotation. The many stops and changes in speed, height and latitude were all carefully noted and taken into account. The final predicted effect, which was a complicated combination of Einstein's predictions for the influence of both speed and gravity on time, was in agreement with the experimental measurements within an accuracy of about 10%. The total cost of the experiment was about $8000 – of which most was spent on air fares.

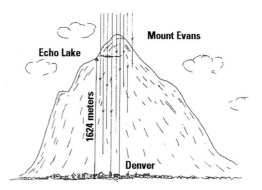

Figure 1.18 Illustration of the decrease in the number of muons on the way from Echo Lake to Denver due to disintegration and absorption in the atmosphere. More muons reached Denver than anticipated because Einstein's time dilation prolongs the life of the unstable muons, particularly those of very high energy. [Courtesy Elfriede Gamow.]

Figure 1.17 The Austrian physicist, Victor Hess (1883–1964) in the balloon basket after one of the early flights on which cosmic rays were discovered. Hess shared the 1936 Nobel prize for physics with the American Carl Anderson, who, in 1932, discovered the anti-electron – the positron – in cosmic rays.

Since this classic experiment of Hafele and Keating, several new experiments involving atomic clocks have been performed which are considerably more accurate. One such experiment was organized by Carroll Alley using US naval flights high over Chesapeake Bay. NASA also sponsored an experiment carried out by Robert Vessot and Martin Levine, using an atomic clock on board a rocket. The rocket reached a height of 6000 miles, and the experiment was able to confirm Einstein's prediction for the effect of gravity on time to an accuracy of within one part in 10 000.

In fact, verification of Einstein's prediction for the retardation of time with speed did not have to wait for the development of atomic clocks. One of the important early tests of special relativity and time dilation involved a 'cosmic ray' particle called a 'muon'. Cosmic rays have their origin in outer space and were first discovered by Victor Hess during a series of rather dangerous balloon flights. He found that, at about a mile above the Earth's surface, the number of energetic particles increased. This implied that the source of this radiation did not come from the Earth itself. Further flights at night and during an eclipse ruled out the Sun as the main source of 'cosmic rays' and implied that they originated outside the solar system.

Muons are among the most penetrating of cosmic ray particles and can still be detected in deep mines, after traversing a huge thickness of rock. This 'weakness' of interaction of muons with matter was very surprising, and it was only after a period of considerable confusion that it was realized that they behave like heavy electrons. The muon is around 200 times heavier than an electron, and although it can now be comfortably accommodated in our present theory of elementary particles its existence is still, in some sense, a mystery. On hearing of the discovery of the muon, the famous physicist Isidore Rabi remarked 'Who ordered that?', and Nobel Prize winner Richard Feynman once wrote on his blackboard 'Why does the muon weigh?' Although physicists may not be too sure why they have the muon, they can nevertheless learn about all its properties by performing careful experiments.

In 1940, Bruno Rossi and his colleagues performed an experiment to measure the intensity of muons – the number arriving per second – both at Echo Lake, near the top of Mount Evans, and down at 'ground level' at nearby Denver. A slab of iron was introduced in the Echo Lake experiment in order to compensate for the anticipated extra absorption of muons that would take place when travelling the extra distance through the atmosphere to the lower site at Denver. The results were compared, and it was seen that measurements at the Denver site consistently showed too few muons compared with the Echo Lake measurements. Rossi consulted Enrico Fermi, his fellow Italian physicist and refugee from Mussolini's Italy. Fermi's solution to the puzzle was to suggest that muons were 'unstable' particles and that some of them underwent a radioactive decay in flight. This could account for the fact that fewer muons than expected reached ground level. Subsequent experiments on muons brought to rest in absorbing material showed that muons were unstable and had a 'lifetime' of just over 2 microseconds. This immediately led to another question: how can such short-lived particles exist in cosmic rays? In fact, in another experiment Rossi also helped determine that most 'primary' cosmic rays are not muons but very energetic protons. Muons are formed from the fragments created when these primary rays shatter nuclei in the upper atmosphere.

What does this experiment have to do with relativity? Well, since cosmic ray muons move at substantial fractions of light speed, we should be able to see the effects of time dilation. According to Einstein, compared with muons sitting at rest, moving muons should live longer. The more energetic the muon, the longer it should live. This was exactly the effect that Bruno

Figure 1.19 This photographic emulsion shows an incoming proton (top right, P) shattering a nucleus. One of the fragments breaks up, emitting a charged pion (π). The pion subsequently decays, forming a muon (μ), which itself decays into an electron. Only the tracks of charged particles are made visible by this method, and several unseen neutral particles (neutrinos) are also involved in these disintegrations. [Courtesy Elfriede Gamow.]

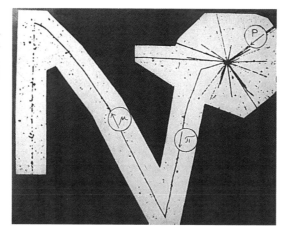

Figure 1.20 The muon storage ring at CERN, the European Laboratory for Particle Physics. A ring of electromagnets cause muons, circulating at close to the speed of light, to follow a circular path 46 feet in diameter. Without the boost to their longevity given by relativistic time dilation, the muons would only be able to orbit the ring about fifteen times before they decayed. In the experiment, muons are observed to orbit hundreds of times before disintegrating. This confirms the relativistic time 'dilation' effect to about 1%. [CERN.]

Satellite navigation systems

Modern satellite navigation systems provide an interesting application of relativity. The Navstar Global Positioning System – known as GPS – is operated by the US Department of Defence. The system consists of twenty-four satellites orbiting the Earth twice every 23 hours and 56 minutes – the length of the so-called 'sidereal day'. The satellites broadcast time signals from an atomic clock on board together with an identification sign. Down below, on the Earth's surface or in an aircraft, receiving equipment picks up the time signals from four satellites and compares them with the readings given by a quartz clock in the receiver. From the four time delays it is possible to calculate the exact position of the receiver. The US military can use the GPS navigation system to pinpoint the position of aircraft, warships and missiles to an accuracy of less than a few metres. In the Gulf war, tank commanders in the desert relied almost totally on such GPS technology to navigate and to fix positions of fuel dumps and so on. Commercial receivers are now available for small boats, cars or even technologically minded hikers. Using very sophisticated processing techniques to analyse the signals, it is possible to obtain position fixes to accuracies of only a few millimetres.

The remarkable precision of the GPS navigation system is a reflection of the accuracy of modern atomic clocks. For example, a clock error of only one-hundred-thousand-millionth of a second would result in a position error of around 30 metres. The use of atomic clocks enables GPS to do much better than this. However, because of this very high accuracy in measurement of time delays, it is necessary to include the predicted corrections due to relativity in the position calculations. There are several relativistic effects that come into play. First, there is the rapid orbital motion of the clock on the satellite, which slows the on-board clock by about one part in 1000 million compared with its Earth-based counterpart. Secondly, the reduced gravity at the satellite's orbital height causes the orbiting clocks to run faster by nearly the same amount. These two effects are roughly the same size and tend to cancel each other out – but not completely. On Earth, there are other smaller relativistic effects which, although less significant, can accumulate and cause inaccuracies if they are ignored. One obvious correction is due to the fact that the rotation of the Earth causes different parts of the globe to move at different speeds. Another correction that may be important is if the clocks in the receivers are themselves moving at high speed, as in a jet aircraft.

Finally, but of fundamental importance, is the fact that the whole of the GPS navigation system is based on calculating time delays using the speed of radio waves from the satellite. Radio waves, as we shall see in the next chapter, are just a part of the spectrum of so-called 'electromagnetic waves'. Light is also an electromagnetic wave: light and radio waves travel at the same speed. It is crucial for the entire basis of the GPS position calculations that the speed of the radio wave signal – the speed of light – is unaffected either by the motion of the satellite in orbit, by the rotation of the Earth, by the motion of the receiver on the Earth, or even by the motion of the Earth round the Sun. The fact that the GPS navigation system works so well is a testament to the truth of this fundamental postulate of relativity.

Advertisement for a satellite navigation receiver, useful for applications such as: gun positioning, astronomy, forestry perimeter marking, seismic sensor positioning and yacht racing. [Navstar Systems Ltd.]

Rossi and his team observed. The discrepancy in the number of muons between the two sites was least for the most energetic muons and exactly matched the expectations of Einstein's time dilation prediction. The effect of time dilation can also be seen more directly. In the absence of this 'Einstein longevity factor' muons would only be able to travel some 600 metres or so in the atmosphere before disintegrating. Contrast this with the fact that muons are created at a height of roughly 20 kilometres up in the atmosphere yet many manage to reach the surface of the Earth. Thus muons in cosmic rays provide very direct evidence for the reality of Einstein's prediction that moving clocks run slow.

We have simplified the discussion of unstable particles somewhat and talked about the muon's 'lifetime' as being about 2 microseconds. This is an average number calculated from the unpredictable decays of many such particles. Nowadays, powerful particle accelerators allow us to produce beams of fast-moving unstable particles like the muon. It is only because of the time dilation effect that physicists are able to do experiments with such beams. A famous experiment at CERN in Geneva stored muons moving at 0.9994 times the speed of light in a magnetic storage ring. By looking for the electrons produced when the muons disintegrated the physicists were able to show that the lifetime of the moving muons was almost thirty times longer than for muons at rest, in precise agreement with Einstein's prediction. In this way, accelerator laboratories around the world daily verify the theory of special relativity.

2   The nature of light

The factor which finally succeeded, after long hesitation, to bring the physicists slowly around to give up the faith in the possibility that all physics could be founded upon Newton's mechanics, was the electrodynamics of Faraday and Maxwell.

Albert Einstein, *Autobiographical notes,* 1949

Fields of force

That gravity should be inate, inherent, and essential to matter so that one body may act upon another at a distance through a vacuum and without the mediation of anything else…is to me so great an absurdity that I believe that no man who has in philosophical matters a competent faculty of thinking can ever fall into it.

Isaac Newton, letter to Richard Bentley

Three pictures hung on the wall of Einstein's study – portraits of Isaac Newton, Michael Faraday and James Clerk Maxwell. These three physicists provided the inspiration for Einstein's great works. With the tools provided by Faraday and Maxwell, Einstein eventually overturned Newton's conception of the universe and the very fabric of space and time which had proved itself so successful for over 200 years.

Newton pictured atoms of matter as having various powers of attraction and repulsion, gravity being the most famous such property. In Newton's scheme of things, the Earth attracts the Moon and all other bodies – such as the famous apple – by virtue of its mass. Newton discovered how this attractive force diminished with distance and, by applying his force law to the Sun and the planets of the solar system, he was able to explain the orbits of the planets. It therefore seems as if the gravitational force is able to exert its influence over vast distances of empty space: this ability is known as 'action at a distance'. As can be seen from his private letters, Newton himself

Figure 2.1 This engraving, made shortly after Newton's death, expresses the awe with which his achievements were regarded. The mood of adulation was captured by Alexander Pope's famous couplet: 'Nature and Nature's laws lay hid in night: God said "Let Newton be", and all was light.' [Courtesy the Museum of the History of Science, Oxford.]

was uncomfortable with this picture, believing that something must pass between the bodies to produce this gravitational attraction. Nonetheless, it was undeniable that the 'action at a distance' theory of gravity was spectacularly successful. This success diverted attention from Newton's more intuitive concerns about such a theory.

Forces acting at a distance also seemed necessary in the early explanations of electricity and magnetism. Today, a wide range of electrical and magnetic effects are familiar features of our modern technological world.

Figure 2.2 Michael Faraday (1791–1867) was the son of a black-smith. As a young man he became apprenticed to a London book-seller and bookbinder. Reading some of the books that came his way led him to become captivated by science. In 1812 he attended a lecture at the Royal Institution given by Humphry Davy. Some time later Faraday asked Davy for a position and showed him the bound notes he had made of his lecture. Davy was impressed, and soon Faraday had a job as his laboratory assistant. Thus began the career of one of the greatest scientists who ever lived. [By courtesy of the Royal Institution.]

Despite such familiarity, children must surely experience some of the fascination of the early scientists when they first encounter magnets. Einstein, in his autobiography, recalls the wonder he felt at the age of four or five when shown a compass by his father. He was amazed that the compass needle should behave in such a determined way, seemingly with a will of its own. When the compass was moved, the needle returned to point in a fixed direction – magnetic north – without anyone touching it. This is the essence of the same sort of 'action at a distance' problem that troubled Newton. Most scientists were prepared to accept that such forces, mysteriously able to act at a distance, were apparently part of the very fabric of the universe – that is until Michael Faraday invented a new way of describing the action of forces in the 1840s.

The fundamental discoveries made by Faraday who was one of the greatest experimental scientists who ever lived, made possible the modern electrical industry. Despite his lack of formal training in mathematics, he seemed to have an incredibly intuitive grasp of the way Nature worked, and he used this to great effect when he performed his experimental investigations. His experiments paved the way for our present understanding of electricity and magnetism, introducing the twin concepts of electric and magnetic fields, the strength of which at any place may be represented by the density of electric and magnetic 'lines of force'.

Faraday's ideas about 'lines of force' can be illustrated by a well-known experiment involving iron filings and a bar magnet. Iron filings sprinkled on a card placed on top of a bar magnet arrange themselves in a pattern that reveals Faraday's magnetic 'lines of force' (see Figure 2.3). Faraday reasoned that such magnetic forces did not just appear when another magnetic body is placed near a magnet. Rather, he thought that the magnet carried round with it a 'field' of magnetic energy distributed throughout the surrounding space. The direction of the magnetic force exerted is indicated by the direction of the lines of iron filings, and the magnitude of the force is represented by the closeness, or density, of the lines. The behaviour of Einstein's compass needle is now explained by the presence of the field lines of the Earth's magnetic field. Action at a distance has been replaced by local interaction, with a force 'field' filling the space.

Other forces, such as electricity, can be visualized in the same way as the magnetic force. Indeed, Faraday saw fields as the ultimate reality, and he believed that light and radiated heat could be explained as 'vibrations' of lines of force. He was also convinced that gravity was analogous to electricity and magnetism, in that there existed gravitational lines of force, and in this belief he was remarkably ahead of his time. Faraday was reputedly an inspiring lecturer who is known to have impressed many of his peers, including Charles Dickens, with the brilliance and lucidity of his talks. His deepest insights and speculations were often communicated through these lectures, especially when called upon to give an impromptu performance. In 1846, his friend and fellow scientist Charles Wheatstone was scheduled to speak at the now famous Royal Institution. At the last minute, Wheatstone panicked and fled, leaving Faraday, as chairman, to cope as best he could. Faraday gave a

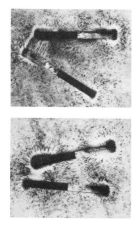

Figure 2.3 The magnetic field between two magnets as revealed by iron filings. The magnets are orientated in opposite directions at the top and in the same direction at the bottom. [Courtesy Elfriede Gamow.]

hastily prepared lecture entitled 'Thoughts on ray vibrations'. Almost twenty years later, the Scottish physicist James Clerk Maxwell attributed the ideas which eventually formed the basis for his electromagnetic theory of light to this unscheduled lecture.

It was Maxwell who transformed Faraday's ideas into the theory of electromagnetism that we know today. 'Maxwell's equations' (see the box on p. 28) are so much a part of every physicist's education that you can buy T-shirts decorated with them. At school in Edinburgh, strangely enough, Maxwell was known as 'Dafty', perhaps because of his home-designed clothes, his country accent and his good sense of humour. However, it is difficult to imagine a more inappropriate nickname for one of the most able theoretical physicists who ever lived. Einstein remarked that 'the pair Faraday–Maxwell has a remarkable inner similarity with the pair Galileo–Newton – the former of the pair grasping the relations intuitively, and the second one formulating those relations exactly and applying them quantitatively'. Maxwell began his transformation of Faraday's ideas into mathematical language using the idea of a hypothetical substance – the 'aether' – which was supposed to fill all of space. Just as sound waves are pressure vibrations of

Figure 2.4 A cartoon from an 1881 edition of *Punch*. It alludes to Faraday's famous response when asked about the usefulness of electricity: 'What good is a newborn baby?' [*Punch*, June 25th, 1881.]

"WHAT WILL HE GROW TO?"

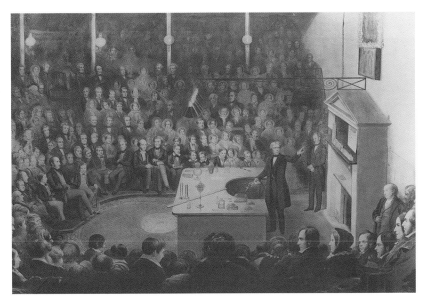

Figure 2.5 Faraday lecturing at the Royal Institution, where he succeeded Davy as director. He was a great popularizer of science, and the Christmas lectures for children which he began still continue today. Faraday belonged to a strict religious sect and declined all public honours and wealth. Nevertheless, he was a Victorian hero, who, like Einstein, was the subject of many admiring biographies. He was approached by the British government during the Crimean War concerning the possibility of making poison gas. He answered that it was possible, but that he would have nothing to do with it. [© Hunterian Art Gallery, University of Glasgow. Painting by Alexander Blaikley.]

Figure 2.6 James Clerk Maxwell (1831–1879) was a prodigy who entered Edinburgh University at the age of sixteen. Maxwell was the first to establish the three-colour model of vision and produced the first colour photograph. He also showed that Saturn's rings could not be solid but must be made up of many small bodies – which he referred to as 'the flight of the brickbats'. Maxwell made fundamental contributions to the molecular model of gases in addition to his mathematical description of electromagnetism. He was the first Professor of the newly erected Cavendish Laboratory in Cambridge. Like Faraday, he was happily married. [Courtesy the Cavendish Laboratory, University of Cambridge.]

the air around us, so Maxwell could represent electric and magnetic fields as distortions in or twists of the aether. In particular, he interpreted light as vibrations of electric and magnetic fields in the aether. His actual mechanical models describing this theory were very elaborate and complicated, but, by using these models as a basis, he was able to develop the four fundamental equations that describe the phenomenon called electromagnetism. He arrived at what are now known as 'Maxwell's equations' in 1864. Only later did he realize that these equations did not depend on the detailed mechanical models from which they had been so laboriously constructed.

It is possible to gain an idea of Maxwell's interpretation of light by analogy with Faraday's field lines. Consider a pair of electric charges. When one charge is moved, a disturbance travels along the field lines, just as a wave travels along a rope when one end is shaken. These electric waves have a characteristic speed predicted by Maxwell's equations. This speed is expressed as a combination of universal constants from magnetic and electrical force laws. When the experimental numbers are substituted into the formula, the speed that comes out is that of light. Light is not just an electric wave, however; it also incorporates a wave of a magnetic field. We can see how this comes about by considering two fundamental discoveries – one made by Faraday and one made by Maxwell.

Faraday discovered that a changing magnetic field creates an electric field. Maxwell had the crucial insight that the converse is also true: a changing electric field generates a magnetic field. Thus, the changing electric field of the wave shown in Figure 2.7 necessarily creates a magnetic field. This changing magnetic field in turn generates an electric field. In this way, a wave of fluctuating electric and magnetic fields travels outwards with the

Maxwell's equations

The four Maxwell equations may be written in the form:

(1) $\nabla \cdot \mathbf{E} = \rho$ (Gauss's law)

(2) $\nabla \wedge \mathbf{E} = -\dfrac{1}{c}\dfrac{\partial \mathbf{B}}{\partial t}$ (Faraday–Lenz laws)

(3) $\nabla \cdot \mathbf{B} = 0$ (no magnetic charges)

(4) $\nabla \wedge \mathbf{B} = \mathbf{j} + \dfrac{1}{c}\dfrac{\partial \mathbf{E}}{\partial t}$ (Ampère's law with Maxwell term)

The symbols $\mathbf{E}$ and $\mathbf{B}$ describe the intensity and direction of the electric and magnetic fields, and the symbol ∇ is the vector differential operator ($\partial/\partial x$, $\partial/\partial y$, $\partial/\partial z$). Standard scalar and vector product notation has been used to summarize these vector equations in an extremely condensed form.

The first equation relates the electric field $\mathbf{E}$ to the electric charge density ρ, and corresponds to Charles Coulomb's inverse square law of force between two electric charges.

The second equation corresponds to Faraday's observations that a changing magnetic field $\mathbf{B}$ creates an electric field, and is the basis for electric motors and dynamos.

The next two equations are the counterparts of these for magnetic fields. Since no free magnetic charges have ever been observed, equation (3) merely states that magnetic charges always come in north and south pole pairs. The final equation has two parts, the first due to the French physicist Andre Marie Ampère and the second due to Maxwell. Ampère's law summarizes the experimental fact that an electric current $\mathbf{j}$ creates a magnetic field, and is the basis of an electromagnet. Maxwell realized that a changing electric field also produces a magnetic field, and it was this insight that led to our understanding of light as an electromagnetic wave. The constant c that appears in the equations is just the speed of light.

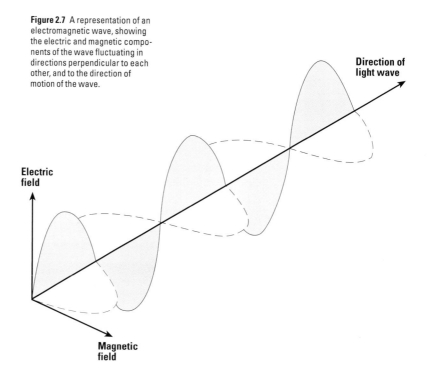

Figure 2.7 A representation of an electromagnetic wave, showing the electric and magnetic components of the wave fluctuating in directions perpendicular to each other, and to the direction of motion of the wave.

Direction of light wave

Electric field

Magnetic field

speed of light. In our example, the distance between the crests of the waves – the *wavelength* of the radiation – depends upon how rapidly the charge is vibrating. Rapid vibration results in waves of short wavelength, whilst slow vibrations produce waves of longer wavelength. Visible light is simply a wave of fluctuating electric and magnetic fields, of wavelength about one-millionth of a metre, with red light having a longer wavelength than blue light. Maxwell realized that wavelengths other than those in the visible range should be possible.

It is hard to imagine modern civilization without radio. It has become part of our everyday existence, to the point that often, in science fiction novels when a spaceship is approaching an unknown planet, the level of radio signal activity is taken as a measure of civilization. Radio waves are actually electromagnetic waves of very long wavelength. They were discovered experimentally by the German physicist Heinrich Hertz, and they provided a dramatic confirmation of Maxwell's equations. Unfortunately Hertz died of blood poisoning at the age of thirty-six and therefore did not live to see Marconi begin to use radio waves as a means of long-distance communication. The whole range of wavelengths – from radio waves, through the infrared and the visible, to ultraviolet, X-rays and gamma rays – form what is now called the *electromagnetic spectrum of radiation*.

Figure 2.8 Heinrich Hertz (1857–1894) was born in Hamburg and was the son of a Jewish lawyer. At university he trained as an engineer, but came under the influence of the great physicist Herman von Helmholtz and changed to physics. Helmholtz encouraged Hertz to enter a prize competition offered by the Berlin Academy of Science for work on electromagnetism. It was this work which led to the discovery of radio waves. [Courtesy the Institution of Electrical Engineers.]

How light behaves

Most of the unfamiliar concepts we encounter in relativity are due to the behaviour of light. In our everyday experience, light appears to be

Figure 2.9 Marconi's signal team at St John's, Newfoundland, struggle in rain and gales to launch a kite-borne receiver for the first transatlantic wireless transmission in December 1901. [Courtesy GEC-Marconi.]

Figure 2.10 A distress call was received from the *Titanic* on April 14th, 1912, by the liner *Carpathia*. Although 50 miles away, the liner responded quickly and was able to save 705 lives. The cartoon honours Marconi. [Courtesy GEC-Marconi.]

transmitted instantaneously because the velocity of light is so large – about 700 million miles/hour. This is to be compared with the speeds that we normally encounter, say in a car (perhaps up to 100 miles/hour) or in an aircraft (maybe 500 miles/hour). This situation with light is in contrast to the behaviour of sound, where we are familiar with effects due to its finite speed of only about 730 miles/hour. A common example is the delay experienced between seeing a lightning flash and hearing a thunderclap during a thunderstorm. Another example is the sonic boom arising from a 'supersonic' aircraft such as Concorde breaking the sound barrier by travelling faster than the speed of sound. In order to convey some feeling about how differently light behaves, it is helpful to compare and contrast its behaviour with that of tennis balls and sound waves in a series of simple 'thought' experiments. We shall consider an experiment consisting of two 'observers' sending a signal to each other, first by throwing a tennis ball, then by sending a pulse of sound (i.e. clapping), and finally by shining a flash of light.

Thought experiment with tennis balls

We shall consider three situations for the two people throwing the ball to each other. First, we look at the simplest situation when both thrower and catcher are standing still; secondly, we consider the case where the catcher is walking towards the thrower who is standing still; and, finally, we look at the case where the thrower is moving towards the catcher as she throws the ball. These three 'experiments' are depicted in the box opposite.

In the first case, the thrower throws the ball towards the receiver with speed v, at, say, 10 metres/second. By measuring the distance between them and the time taken for the ball to reach the catcher, both thrower and receiver measure the same speed, v, as 10 metres/second.

Experiment with tennis balls

Experiment 1

A tennis ball is thrown in a horizontal direction at a speed of 10 metres/second. The ball would actually have to be thrown in a slightly upwards direction, as shown at the top of the figure because gravity causes the ball to travel along a parabolic path (shown as a full line with the ball at the top of the trajectory). This up–down component of the motion is independent of the horizontal motion, so we have ignored this component of the motion in the text. Air resistance has also been eliminated from this 'thought experiment'.

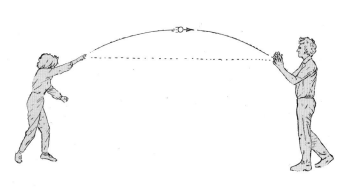

Experiment 2

In this experiment the tennis ball is thrown as before, at 10 metres/second, but this time the catcher runs towards the ball at 5 metres/second. The closing speed between the ball and the catcher is 15 metres/second, just the sum of these two speeds.

Experiment 3

In the third experiment the thrower is now running at 5 metres/second as she throws the ball with a horizontal speed of 10 metres/second. The catcher sees the ball approaching him with a speed of 15 metres/second.

In the second experiment, the catcher is moving towards the thrower at a speed u, which is, say, 5 metres/second. In this case, the catcher starts from further away and measures the *relative* speed of the ball as $v + u$ or 15 metres/second, putting in the numbers of our experiment.

In the third experiment, the thrower is moving towards the receiver with speed u, 5 metres/second, as she throws the ball. The receiver then sees the ball moving towards him with a speed $v + u$, or 15 metres/second, as in our second experiment. In both cases, when either the thrower or the catcher is moving, the relative speed of the ball approaching the catcher is the sum of the two speeds, $v + u$. This is an example of relativity according to Newton.

Thought experiment with sound

In this series of experiments, the 'ball' of the previous examples is replaced by a 'source' of sound: the experimenter loudly clapping her hands once. This is illustrated in the box opposite. A sound wave consists of a series of pressure variations caused by the molecules of the air in which the wave is moving being alternately bunched up and spaced out. The speed of sound therefore depends only on properties of the air, such as its pressure and density.

When the source of the sound and the listener, or 'receiver', are both standing still, the time the sound takes to travel between them gives a measurement of the speed of sound in air, which we shall denote by the symbol v, as for the ball example. In this case, of course, v is equal to some 330 metres/second – the speed of sound in air.

In the second experiment, the listener is now moving towards the source of sound with speed u, 5 metres/second. He now hears the sound earlier than if he had stayed still. The speed at which the noise appears to travel is therefore the sum $v + u$, as in the tennis ball case. Using the value for the velocity of sound in air, this is 335 metres/second.

The last experiment with sound gives a very different result from that conducted with tennis balls. In this case, the source of the sound, the person who claps her hands, is moving towards the receiver with speed u, 5 metres/second, which, of course, is much less than the speed of sound. However, unlike the case of the tennis ball, where the ball went faster if the thrower was moving in the direction the ball was thrown, the speed of sound through the air is unaffected by the motion of the source. This is because the speed of sound is purely a property of the intervening 'medium' i.e. the air, between the two experimenters. Thus, in this case, the speed measured for the sound to reach the receiver is still just v – the velocity of sound in air, 330 metres/second.

Notice that if we repeat the sound experiment in a vacuum, no sound will travel between the source and receiver, since there are no air molecules to transmit the pressure variations. Physicists of Einstein's time were therefore used to the idea that a wave needed a 'medium' for its transmission. Roughly speaking, they believed that a wave consists of some property (such as pressure or height) changing its value ('wiggling') and that there needed to be something *doing* the 'wiggling' – air molecules for sound waves, water molecules for water waves, and so on. We shall see that it was the assumption

Experiment with sound

Experiment 1

A pulse of sound from a hand clap moves towards the receiver at 330 metres/second, the speed of sound in air.

Experiment 2

In the second experiment with sound, the receiver is running towards the source of sound at 5 metres/second. The closing speed of the pulse of sound is the sum of the two speeds, that is 335 metres/second.

Experiment 3

In the third experiment it is the person who is the source of the sound who is running at 5 metres/second when she claps her hands. Sound is a pressure disturbance in the air and the velocity of the pulse depends only on properties of the air. Motion of the source thus leaves the speed of sound unaltered, at 330 metres/second.

Experiment with light

Experiment 1

A short pulse of light travels between two stationary experimenters. Light, like the tennis ball, falls to Earth, but the speed of light (about 300 000 000 metres/second in a vacuum) is so great that this component of the motion is completely negligible. This thought experiment also ignores the small change in speed due to the air through which the light is travelling.

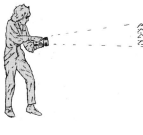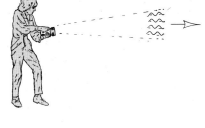

Experiment 2

In this experiment the receiver runs towards the signaller at 5 metres/second. Despite this relative motion, the receiver still observes the light to be approaching at the same speed as before, i.e. 300 000 000 metres/second. The receiver does not see light approaching with a speed which is the sum of the two velocities, or 300 000 005 metres/second. This is unlike either of our two previous experiments with tennis balls and sound waves.

Experiment 3

In this final experiment it is the signaller who is running towards the receiver at 5 metres/second. The receiver still observes the light approach at the same speed as before, 300 000 000 metres/second. This result is similar to that observed for a pulse of sound.

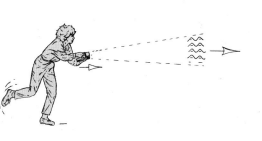

that there must be some medium through which light waves wiggled – the so-called aether – that led to all sorts of confusion around the turn of the century. Einstein's great contribution to physics was to sweep away all the unnecessary complications generated by this mythical aether.

Thought experiment with light

We will now repeat the three experiments with one of the two experimenters using a flashlight to send out a short pulse of light to the other. This is shown in the box on p. 34. In the first case, both experimenters are stationary and they measure the velocity of light v. In fact, because this is so large, 300 000 000 metres/second, they would have difficulty measuring the tiny amount of time taken by the light to travel between them. Since this is a thought experiment, however, we need not worry about such details here; we will look at how such difficult experiments have, in reality, been performed later in this book.

We will now consider what happens when the receiver of the light signal is moving towards the signaller. If the receiver moves towards the light source at speed u, he will, to his surprise, observe the light pulse travelling towards him at the same speed v that he measured when he was standing still! (If he can only move at a speed of 5 metres/second, this would, of course, be difficult to demonstrate convincingly in a real experiment, since the receiver's speed is so small compared with that of light. This was why, in Chapter 1, we used the *Starship Enterprise*, which is able to move at speeds approaching that of light so that the relativistic effects are much more dramatic. Nevertheless, the result that the moving receiver measures the same velocity v as when he was standing still is true and has been checked experimentally, as we shall see.)

What about the last case, when the light source is moving towards the observer with speed u? In this case, exactly as for sound waves, the receiver measures the speed of the light pulse as v, the same speed as he obtained with a stationary source. The same practical experimental difficulties will be true for this experiment; again, however, the result has been confirmed experimentally.

From these three sets of thought experiments, we see not only that light does not behave like tennis balls, but also that it does not behave like sound waves travelling through some intervening medium, such as air. Moreover, in the two cases describing the behaviour of tennis balls and sound waves, we did not need to be very careful in spelling out how we actually measured the different times taken between sender and receiver to calculate the corresponding speeds. We tacitly assumed that in all situations, with both stationary and moving experimenters – there was a 'universal time' on which everyone could agree. As we discussed in chapter 1, experiment shows that this is false. In situations involving speeds close to the velocity of light, we must carefully re-think how each experimenter goes about measuring times and distances. Before we do this, we shall make a historical diversion. After explaining how belief in the 'luminiferous aether' came about, we will describe the most famous attempt to measure an effect caused by it. In the next chapter, we will see how Einstein resolved all the problems caused by the aether by abolishing it altogether!

One small caution is in order. In our experiment with tennis balls, we used everyday, realistic speeds for our tennis balls and arrived at the commonsense result for the addition of velocities. According to Einstein, this Newtonian velocity addition formula is only an approximation, appropriate for low speeds. In our tennis ball example, where all the speeds were much less than the speed of light, adding velocities according to Newton's law is a very good approximation. However, if the tennis balls were replaced by fast missiles, and the source and receiver were replaced by spaceships capable of moving at speeds that are a significant fraction of the speed of light, we would find that the simple velocity addition law due to Newton no longer works. We shall look at this later, together with the experiments that have been performed to test the velocity addition formula due to Einstein.

The search for the aether

As a student I got acquainted with the unaccountable result of the Michelson experiment and then realized intuitively that…[the problem would be solved]…if we recognized the experimental result as a fact. In effect, this is the first route that led me to what is now called the special principles of relativity.
Albert Einstein, *Kyoto Address*, 1922

Einstein thought of the era of Newton as the 'childhood of science,' and it is a good starting point for our story of light and the aether. It was in 1676, when Newton was thirty-four, that a young Danish astronomer, Ole Roemer, discovered that light travels at a finite speed. Roemer was working at the Paris Observatory making painstakingly accurate measurements of the eclipses of the moons of Jupiter. He only wrote one short scientific paper, but its conclusion was striking enough to ensure his place in the history of physics. In the paper he wrote:

> I have been observing the first satellite of Jupiter over eight years. The satellite is eclipsed during each orbit of the planet on entering its vast shadow. I have observed that the intervals between eclipses vary. They are the shortest when the Earth moves towards Jupiter and longest when it moves far away from it. This can only mean that light takes time for transmission through space. The speed of light must be so great that the light coming to us from Jupiter takes 22 minutes longer to reach us at the farthest end of our orbit round the Sun, than at the other end when we are nearest Jupiter. That is to say, light takes about 10 minutes to travel from the sun to the Earth; it does not travel instantaneously as alleged by M Descartes.

Figure 2.11 The Paris Observatory, where Roemer first measured the speed of light.

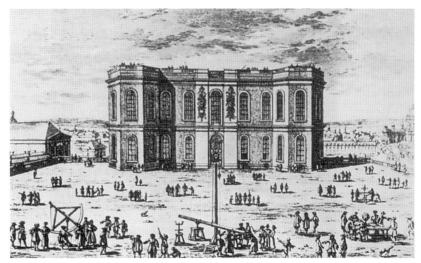

Figure 2.12 The diagram shows an eclipse of Jupiter's moon Io, when the Earth is on the same side of the Sun as Jupiter. Six months later, the Earth has moved to the opposite side of the Sun. Since Jupiter takes almost eleven Earth years to complete an orbit of the Sun, in six months it will appear to have barely moved. With the Earth on the opposite side of the Sun to Jupiter, the eclipse of Io occurs sixteen minutes later than before. This is due to the time taken by light to travel across the orbit of the Earth.

Roemer was apparently content to have shown that the speed of light was not infinite. Despite the fact that it was at the Paris Observatory that the size of the Earth's orbit was first determined, it is not recorded whether Roemer ever divided this distance by the time taken by light to cross this distance to obtain a value for the speed of light. It was apparently left to the Dutch physicist Christian Huyghens two years later, in 1678, to make the first estimate of the speed of light from Roemer's data. He obtained a value of about 200 million metres/second, a speed so much faster than any other encountered on Earth that it was not surprising that Roemer's explanation generated considerable

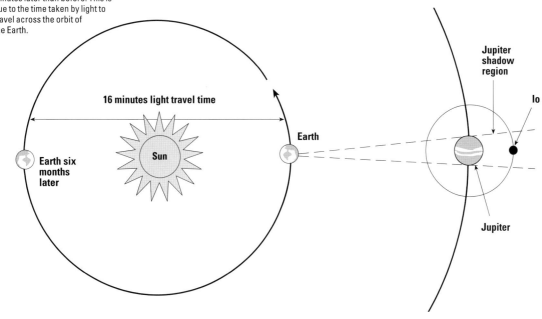

controversy. His conclusion was not even believed by the astronomer Giovanni Cassini who was his director at the Paris Observatory. Despite this scepticism in Paris, Roemer's discovery was greeted with acclaim in England, and he later met both Isaac Newton and Edmund Halley, discoverer of his eponymous comet. Modern measurements of the speed of light made using more accurate methods agree on a value close to 300 000 kilometres/second.

We now move forward in time to 1879 when there occurred renewed interest in the eclipses of Jupiter's moons. Maxwell wrote to David Todd, Director of the Nautical Almanac Office in Washington, to enquire about the feasibility of using these eclipses to measure the velocity of the Earth through the aether. The essence of the idea is the following. First of all, it was assumed that light was a vibration of the aether – in the same way as sound is known to be a vibration of the molecules of the medium through which it is travelling. Thus, if the Earth is moving through the aether, then presumably the velocity of light would be less when the light was moving upstream against the flow of the aether than if it were moving with the flow of the aether. Maxwell calculated that Roemer's time delay would vary by up to 1 second as Jupiter orbited the Sun. Unfortunately for Maxwell, Todd replied that the present astronomical data were not sufficiently accurate for this to be a viable experiment.

In his letter to Todd, Maxwell lamented that it was apparently impossible to perform a terrestrial experiment to detect the Earth's motion through the aether. Maxwell pointed out that, on Earth, the speed of light was measured by returning a beam of light to its starting point using a mirror. This has the consequence that the time gained when the light motion is aided by the aether flow is almost exactly cancelled by the time lost by its battle against the flow on its return journey. Thus, Maxwell thought the effect would be too small to measure by any Earth-based experiment. However, Maxwell's letter was read by a young American called Albert Michelson. He became convinced that Maxwell's 'impossible' experiment could be performed. For his work on this problem, Michelson would later become the first American to be awarded a Nobel Prize in physics.

Albert Michelson entered the US Naval Academy at Annapolis at the age of seventeen. When he graduated in 1873, he became a science instructor for the US Navy. Michelson soon realized that an accurate value for the speed of light was essential for all that he taught about physics and navigation. In 1878, at the age of twenty-five, Michelson performed a classic experiment that gave the most accurate measurement of the speed of light at that time. He obtained the value 299 910 kilometres/second, very close to today's accepted best value. Inevitably, Michelson became both interested in the aether and in attempts to measure any effects due to its presence. The aether was indeed a curious medium. Because the speed of light is much greater than the speed of any other type of mechanical disturbance, and as light is a wave vibration transverse to the direction of motion of the wave, in order to account for this high speed, the aether must provide restoring forces of a magnitude that only normally occur in solids. Thus, all space had to be

Figure 2.13 Hendrik Antoon Lorentz (1853–1928) was, in Einstein's opinion, the most well-rounded and harmonious person he had ever met. He was also a great physicist. At the memorial service after Lorentz's death, Einstein said: 'The enormous significance of his work consisted therein, that it forms the basis for the theory of atoms and for the general and special theories of relativity. The special theory was a more detailed expose of those concepts which are found in Lorentz's research of 1895.' [Courtesy AIP Emilio Segrè Visual Archives, Landé Collection.]

filled with a medium apparently more resistant to transverse vibrations than steel, yet sufficiently insubstantial that the planets could travel through it year after year suffering no detectable resistance! Many ingenious experiments were devised to detect some measurable effect of the aether. None were successful, and this led to equally many imaginative explanations to explain why no effects had been seen. Since the search for the aether turned out in the end to be a wild goose chase, we shall only look at the most famous of these experiments – the so-called Michelson and Morley Experiment.

It was in the early 1880s, whilst on study leave in Germany, that Michelson came across Maxwell's letter to Todd. After much thought, Michelson developed a very sensitive 'interferometer' (see the box on p. 40) – paid for by Alexander Graham Bell – to perform the experiment that Maxwell had suggested. He was very disappointed, however, that he was unable to detect any effects due to the aether. On returning to America, Michelson resigned from the Navy and joined the Case School of Applied Science in Cleveland, Ohio. The null result of Michelson's experiment had been so unexpected that many people, including the great Dutch physicist Hendrik Antoon Lorentz, made great efforts to reconcile it with the aether theory. In 1887, Michelson decided to undertake an improved version of his earlier experiment, and so, with a colleague, chemist Edward Morley from nearby Western Reserve University, he performed the much more precise version of the experiment that is now found in all the textbooks (see the box on p. 40). Much to the disappointment of the experimenters and Lorentz, they saw no detectable effects that could be explained as being due to the aether.

One of the most striking 'explanations' for Michelson and Morley's inability to detect any 'aether drift' was put forward by the Irish physicist George Francis Fitzgerald. He suggested that the entire apparatus shrank! Specifically, Fitzgerald suggested that the apparatus contracted in length along the direction of the Earth's motion through the aether, and that this could account for their null observation. Absurd though it sounds, this idea contains a germ of the truth, as we shall see later. Fitzgerald was refreshingly honest, and once said of himself:

> I am not in the least sensitive to having made mistakes, I rush out with all sorts of crude notions in the hope that they may set others thinking and lead to some advance.

In fact, Lorentz, unaware of Fitzgerald's idea, had independently put forward a similar hypothesis based on an atomic view of matter. When, in 1894, Lorentz heard of the Irish physicist's earlier proposal of the contraction theory, he immediately wrote to Fitzgerald. Rather than worry about who had thought of the idea first and who, ultimately, should receive the credit, Fitzgerald was delighted that Lorentz had come to a similar conclusion. Scientists are not always so generous.

The contraction hypothesis, although it could account for the null result of Michelson and Morley's experiment, did not survive a second aether-drift experiment carried out by Roy Kennedy and Edward Thorndike

The Michelson and Morley Experiment

The famous experiment of Michelson and Morley attempted to measure an effect due to the motion of the Earth through the aether. In order to understand the principle involved, consider the following nautical analogy. Imagine a race between two identical boats on a river. In our analogy, the flow of the river represents the motion of the Earth through the aether and the voyages of the two boats represent the different paths of light beams through the two arms of the apparatus. One boat crosses the river and returns to its starting point: the other covers the same distance first down-river, aided by the current of the river, and then upstream back to the starting point, moving against the current. Which boat will win? The time gained by the boat travelling with the current is almost cancelled out by the time lost when travelling against the flow. On the other hand, the boat travelling across the river has to aim slightly upstream to avoid being swept downstream by the current. A similar situation holds for the return journey. A simple calculation shows that the boat travelling across the river will always win.

The light paths in the Michelson and Morley apparatus shown in the illustration are precisely analogous to the two courses of our boat race on the river. Flag B on the river corresponds to the beam-splitter – a half silvered mirror – which is also labelled B. The flow of the aether is assumed to be parallel to a line from the beam-splitter to the mirror labelled C. The path from B to A and back is the same distance as the path from B to C and back. A source of light sends a beam towards the apparatus, where it is split by the half silvered mirror. The two halves of the beam are reflected by the mirrors at A and C, recombine back at the beam-splitter and then travel on towards the telescope. Any

difference in time taken by the light in travelling the two paths will result in the light waves of the two beams being no longer exactly in step. Two such misaligned waves will 'interfere' with each other and cause characteristic interference fringes – bands of high- and low-intensity light.

Michelson and Morley did not know the direction of the 'aether wind'. Their apparatus was mounted on a massive stone slab floating in mercury. By rotating their apparatus they were able to change the direction of the supposed aether wind relative to the arms of the apparatus. Michelson and Morley reluctantly concluded that they saw no sign of

the expected shift in the interference fringes. They also made observations day and night, as the Earth spins on its axis, and in all seasons of the year, as the Earth moves round the Sun. No fringe shift was observed. Michelson's first paper concluded with the statement: 'The result of the hypothesis of a stationary aether is thus shown to be incorrect.' Our simplified description of the Michelson and Morley experiment has ignored many practical details and precautions that were taken into account in the real experiment. The conclusion remains the same: there was no detectable sign of the aether.

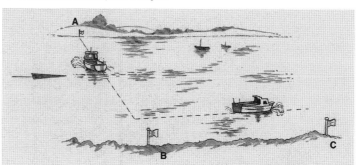

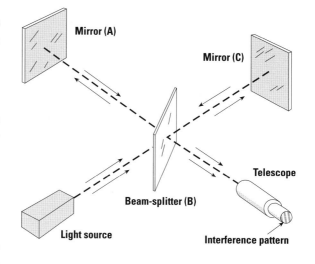

Mirror (A)

Mirror (C)

Telescope

Beam-splitter (B)

Light source

Interference pattern

in 1932. Their apparatus was a variant of that of Michelson and Morley, with different length arms and the whole construction fixed in place as the Earth orbits the Sun. Kennedy and Thorndike saw no shift of the interference fringes as days and months went by, and this null result could not be explained by the Lorentz–Fitzgerald contraction. Indeed, the contraction hypothesis was in trouble even before this final nail in the coffin for the aether. Lorentz had arrived at his contraction prediction in terms of a very elaborate attempt at constructing an electron theory of matter. This theory predicted other effects, besides the contraction, which were not reproduced experimentally. Thus, nineteenth century physics had no satisfactory explanation for the null result of Michelson and Morley. The concept of the aether was so deeply ingrained into the thinking of the time that the failure of all attempts to 'save the aether' came as a real blow to the physics community.

Figure 2.14 George Fitzgerald (1851–1901) was an early flying enthusiast. Notice that Fitzgerald did not abandon his top hat for this experiment. [Courtesy Trinity College, Dublin.]

Is Cleveland, Ohio, the centre of the universe?

By 1894, the aether theory was in a desperate state. Robert Cecil, Lord Salisbury, addressed the British Association and admitted that

> For more than two generations the main, if not the only, function of the word 'aether' has been to furnish a nominative case to the verb 'to undulate'.

The aether also defined what was called a 'preferred reference frame': within this reference frame, all other motions could be described relative to the stationary aether. Michelson concluded his paper that described the results of his first experiment with the following words: 'The result of the hypothesis of a stationary aether is thus shown to be incorrect.' In other words, either Cleveland, Ohio, where Michelson had performed his experiment, was the centre of the universe, around which everything else revolved – a hypothesis not many people were prepared to consider – or it was not possible to detect any motion relative to any proposed aether.

Several people came close to discovering relativity. Lorentz had tried to incorporate his electron theory into Maxwell's equations and had been forced to introduce a new parameter that he called 'local time'. However, neither he nor anyone else dared to suggest that time itself was not an absolute quantity. Similarly, the great French mathematician and physicist, Henri Poincaré, developed Lorentz's work into a beautiful mathematical framework. He, too, was very close to describing relativity. In 1904, Poincaré concluded a lecture with the words:

> Perhaps we must construct a new mechanics of which we can only catch a glimpse… in which the velocity of light becomes an unpassable limit.

Lorentz and Poincaré had discovered the formal mathematical structure of relativity, a fact that Einstein acknowledged two months before he died with the words:

> Lorentz had recognized that the transformations named after him are essential for the analysis of Maxwell's equations, and Poincaré deepened this insight still further.

Both Lorentz and Poincaré, however, were working purely in the context of electromagnetism: it took Einstein to realize the full implications of Michelson and Morley's famous null experiment.

In retrospect, it is clear that what was needed was someone who could see that the message of the Michelson and Morley Experiment was not 'Go away and devise more elaborate theories of the aether', but rather 'The aether does not exist'. It was not until eighteen years after this experiment that such an individual arrived. In 1905, the young Einstein was not only bold enough to countenance this heresy, but also tenacious enough to follow through the implications of this manifest experimental fact, even if it meant changing the very concept of time. Curiously, despite the quotation on p. 36, Einstein does not refer to the Michelson and Morley Experiment in his famous paper, and, at various times, he even denied knowledge of it prior to his 1905 paper. In his marvellous scientific biography of Einstein, *Subtle is the Lord*, Abraham Pais details all these contradictory statements of Einstein, and speculates as to why Einstein was clearly reluctant to acknowledge the possible influence of the Michelson and Morley Experiment on his thinking.

An aethereal footnote

I have for long thought that if I had the opportunity to teach this subject, I would emphasize the continuity with earlier ideas. Usually it is the discontinuity which is stressed, the radical break with more primitive notions of space and time. Often the result is to destroy the confidence of the student in perfectly sound and useful concepts already acquired!
John Bell, in *Speakable and Unspeakable in Quantum Mechanics*

As we can see from the above quote, the Irish physicist John Bell, best known for his fundamental contributions to quantum mechanics, firmly believed that relativity was best introduced to students using an approach that emphasized continuity with earlier ideas – rather than forcing upon them the radical break advocated by Einstein. Bell took up Lorentz's ideas, using Maxwell's equations applied to moving charges to justify the length contraction first proposed by Fitzgerald (see Figures 2.15 and 2.16). However, as we have said, this contraction alone does not account for the null result of Kennedy and Thorndike's later variant of the Michelson and Morley Experiment. Bell therefore also showed that an orbiting electron not only has its stationary circular orbit squashed to an ellipse when in motion, but also that the time for completion of one orbit is increased or 'dilated' by just the factor predicted by relativity. With this 'Larmor time dilation' included (it was first calculated by Joseph Larmor in 1900), the non-detectability of motion with respect to the aether can be explained. Since it is experimentally impossible to say which of the two systems is 'really at rest', it now becomes a matter of philosophy whether one prefers to retain the idea of a state of 'real' rest defined by an aether.

Einstein declared the notions 'really resting' and 'really moving' to be meaningless: only the relative motion of two or more objects had any reality. Bell preferred to accept a 'Lorentzian pedagogy' to explain the results of Einstein's theory, without accepting Lorentz's philosophy. It has to be said, however, that no one has yet succeeded in building a fully adequate and consistent theory of matter along the lines suggested by Lorentz. Bell accepts this and retreats to the position that the laws of physics must be the same for moving and stationary observers when account is taken of the Lorentz–

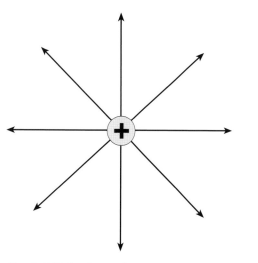
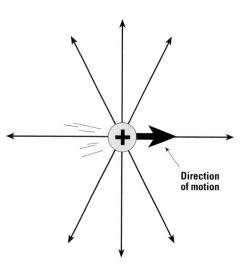

Electric field of stationary charge

Electric field of rapidly moving charge

**Direction
of motion**

Figure 2.15 The electric field of a stationary charge (left) and of a charge moving rapidly (right). The field lines of the moving charge are compressed along the direction of motion. A magnetic field (not shown), which circulates around the direction of motion, is also created by the moving charge.

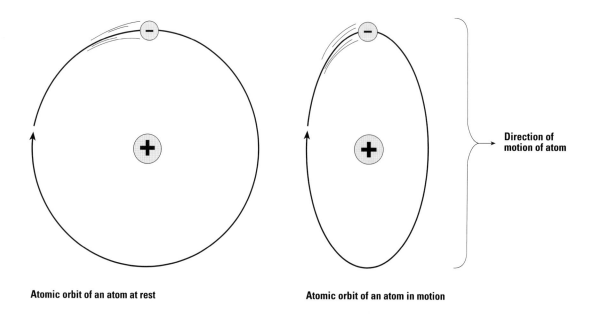

Atomic orbit of an atom at rest

Atomic orbit of an atom in motion

**Direction of
motion of atom**

Figure 2.16 A circular orbit of an electron in an atom (left) and the squashed orbit of the same atom in motion (right). This squashing is the result of the moving electric and magnetic fields, together with a modification of the electron momentum proposed by Lorentz. Using these results, the orbit becomes squashed by the Lorentz–Fitzgerald factor, while the orbital period is lengthened by the 'time dilation factor' (*see the Appendix for a derivation*).

Fitzgerald contraction and Larmor time dilation. In modern technical language, we would say that the laws of physics must be 'Lorentz-invariant'. This was the great clue for Einstein. Maxwell's equations were indeed the same for moving and stationary observers; Newton's laws were not. Einstein was bold enough to believe Maxwell and to question Newton.

It is therefore true that one can, to some extent, motivate and make plausible both the Lorentz–Fitzgerald contraction and the Larmor time dilation by means of a much over-simplified model of matter. However, the difficulties in extending these arguments to any realistic description of matter seem so great that, for us and for the great majority of physicists, it seems preferable to follow Einstein. Einstein starts from the premise that only *relative* motion is meaningful, and conjectures that the laws of physics look the same for all observers moving with uniform speeds. There is no preferred 'stationary' observer and consequently no need to introduce an aether.

3   Light and time

The introduction of a luminiferous aether will prove to be superfluous.

Albert Einstein, in 'The electrodynamics of moving bodies', 1905

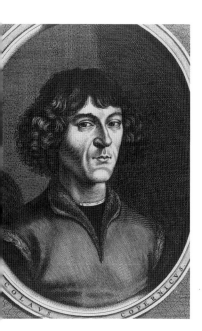

Figure 3.1 Nicolaus Copernicus (1473–1543) spent most of his life as a canon at Frauenberg Cathedral. His idea of placing the Sun at the centre of the system of planets was the crucial first step that inaugurated the modern scientific revolution. The second step was taken by the German astronomer, Johannes Kepler (1571–1630), who proposed that the planets move in elliptic, rather than circular, orbits. With Kepler's innovations, the Copernican system was significantly more accurate in astronomical prediction than the rival Earth-centred view. These developments occurred just before the telescope transformed our picture of the heavens. [Courtesy AIP Emilio Segrè Visual Archives.]

The momentous day in May

In May of 1905, Einstein was twenty-six years old, and his ten-year struggle with the problems of relativity was about to come to a triumphant climax. About a year before this, he had begun to feel that the velocity of light must be universal – independent of the motion of the source. If this were true, then there was no need to worry about motion relative to any mythical aether, and the null result of Michelson and Morley became obvious: the speed of light is the same in both arms of the apparatus, whatever direction they are pointing relative to the Earth's motion. But the Earth does move round the Sun – so something was wrong with the 'relativity' of Galileo and Newton and their familiar addition of velocities, at least where light is concerned. As we asserted in chapter 2, and as we shall show in the next chapter, in Einstein's relativity speeds do not add up in the expected way. We are also forced to re-think our notions of space and time. This new vision of space and time is what we shall look at in this chapter. Let us start by recalling what Galileo and Newton believed, before looking at Einstein's version of the relativity principle.

In the sixteenth century, it seemed natural to believe that, if the Earth was moving, neither an arrow shot straight up nor a stone dropped from a tower would follow the same straight-line path. It seemed 'obvious' that the arrow and the stone would land either ahead or behind the archer and thrower, depending on which way the Earth was moving. Consequently, the idea of a moving Earth was almost unthinkable: birds, even the air and the oceans, would be left behind in its wake! How can this argument be coun-

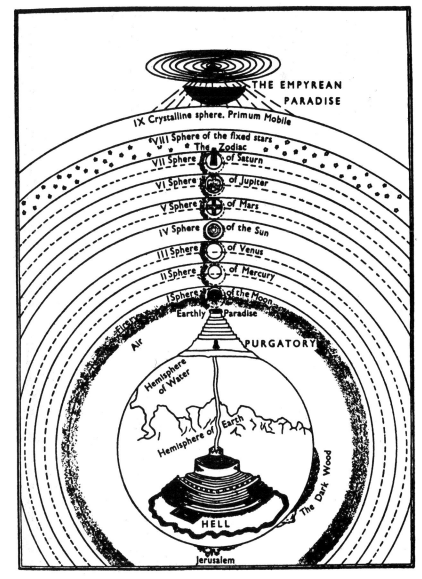

Figure 3.2 Dante's vision of the Medieval universe as depicted in *The Divine Comedy* had Hell at the centre of the Earth and the cosmos. Paradise was up beyond the system of planets. Life on Earth is precariously balanced between these two possible final fates, both geographically and morally.

tered? Isaac Newton credits Galileo with the discovery of the principle of relativity. Galileo Galilei was a forceful Italian scientist, probably best remembered for his fight with the Catholic Church and the Inquisition. In order to understand how such a conflict between Science and the Church was possible, one has only to look at Dante's vision of the Medieval universe (Figure 3.2), in which Earth was firmly fixed at the centre. The Copernican picture, with the Earth whirling round the Sun, was still portrayed as only a convenient device for performing complicated astronomical calculations. To insist that the Earth was moving round the Sun – in an obviously subordinate role –

clearly questioned some very fundamental beliefs. Opposition was not con-fined to Rome: on hearing a rumour about the work of Nicolaus Copernicus, Martin Luther remarked:

> People gave ear to an upstart astrologer who strove to show that the Earth revolves, not the Sun and the Moon. This fool wishes to reverse the entire science of astronomy.

In Galileo's day, trains and planes did not exist, and ships probably provided the smoothest and least jarring means of travel. Galileo imagined shutting himself up in the cabin of a ship with only some birds and a bowl of water with some fish in it for company. Once inside this enclosed cabin, Galileo asserted that, by observing only the motions of the fish and the birds, it would be impossible to tell whether the ship was at rest or moving with a steady speed. Obviously, one could detect motion if the ship was swaying from side to side or lurching up and down, but, if it was sailing on a calm sea in 'uniform motion', it would be impossible to tell without looking outside. Nowadays, we have many fairly smooth forms of transport, and this 'relativi-ty principle' is second nature to us. Most readers will have experienced a momentary confusion on a railway journey when they are uncertain as to whether *their* train or the *neighbouring* train is moving out of a station.

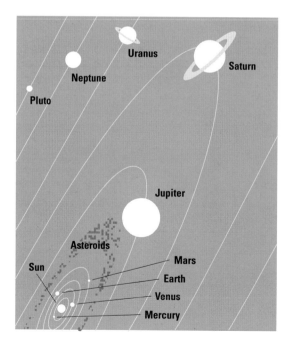

Figure 3.3 A modern illustration of the Copernican model of the plane-tary system. This illustration includes the planets Uranus, Neptune and Pluto, which were unknown to Copernicus. (Not to scale.)

Newton was first to give a precise mathematical formulation of Galileo's ideas. It was also Newton who first gave clear meaning to the idea of a 'force': that all bodies continue in a state of rest or uniform motion unless acted on by an external force.

Two centuries later, Einstein developed the ideas of relative motion. Einstein's 1905 paper on relativity was unusual for many reasons. It contained no reference to any other scientific paper; it was written in a direct and clear style; and it could be understood without any advanced mathematics. It had, and still has, an extraordinary logical clarity that gives the somewhat misleading impression that his theory was the effortless product of pure thought. The merest hint of the human side of Einstein is given by his acknowledgement in the paper of the 'loyal assistance of my friend M. Besso'. Einstein had been frustrated for over a year by his inability to reconcile his guess about the constancy of the speed of light and the usual 'Galilean' addition of velocities. Then, 'one momentous day in May', he visited his friend and fellow patent clerk, Michelangelo Besso. Einstein recalled starting the conversation with him in the following way: 'Recently I have been working on a difficult problem. Today I come here to battle against that problem with you.' After they had discussed every aspect of the problem, Einstein suddenly understood where the key to the problem lay. The following day, he went back to Besso and, without even saying hello, he said, 'Thank you. I've com-

Figure 3.4 A painting of Galileo's trial at which he was forced to recant his Copernican opinions. The Catholic Church declared that his heresies were 'more scandalous, more detestable, and more pernicious to Christianity than any contained in the books of Calvin, of Luther, and of all other heretics put together.' [BAL2344 *Trial of Galileo*, 1632 by Anonymous, Private Collection/The Bridgeman Art Library, London.]

pletely solved the problem.' As he said long afterwards, 'At last it came to me that time was suspect.'

David Hilbert, the great German mathematician, once explained Einstein's contribution with the words:

> Do you know why Einstein said the most original and profound things about space and time that have been said in our generation? Because he had learned nothing of the philosophy and mathematics of space and time!

After this great insight about the nature of time – that there is no absolute time that is the same for everyone, everywhere, moving at any speed – it took Einstein only five weeks to complete and write up his paper for publication.

The logical perfection of Einstein's finished paper gives no hint of his years of struggle or of the puzzling experimental results which motivated his search. One controversy that still rumbles on is whether or not Einstein knew of the results of the Michelson and Morley Experiment. There is no reference to the experiment in his paper, and at various times during his life Einstein claimed he only learnt of the experiment after his 1905 paper was published. However, in a conversation in 1952 Einstein was not so sure. He told the US physicist Abraham Pais, that he realized

> that he had also been conscious of Michelson's result before 1905, partly through his reading of the papers of Lorentz, and more because he had assumed this result of Michelson to be true.

This seems the most likely explanation. In his book *Subtle is the Lord*, Pais speculates that Einstein's discovery

> was so overwhelming that it seared his mind and blotted out reflections and information that had been with him earlier, as the result of deep-seated desires to come closer to the divine form of pure creation.

It is a strange sort of revolution that goes unnoticed at first. Einstein's sister, Maja, has described how her brother waited anxiously for a reaction to his now famous paper. The paper was titled, perhaps somewhat obscurely, 'The electrodynamics of moving bodies'. Despite Einstein's great hopes, his paper was not greeted by a barrage of either praise or criticism; instead, there was almost total silence! The next few issues of the journal *Annalen der Physik* made no reference to his paper. It was therefore not surprising that Einstein was overjoyed when he eventually received a letter from one of the great men of German physics, Max Planck, originator of the quantum theory. Planck, in fact, was asking for clarification of some points in the paper that he found obscure, but Einstein was delighted that his paper had been read by one of the most famous physicists of the time. Planck and Einstein held a mutual admiration for each other throughout their lives. It was Planck who encouraged Einstein to take up a professorship in Berlin – which eventually led to Einstein renewing the German citizenship that he had renounced on his arrival in Switzerland. He later remembered this renewal of citizenship as 'one of the follies of his life'.

Time is relative

We have stated several times that Einstein's relativity requires us to revise our ideas of an 'absolute time' – the notion that there is one time that is the same for everyone, anywhere, irrespective of their relative speeds. We can see how the need to modify our ideas about time comes about directly from the two assumptions in Einstein's paper. His first assumption concerned the speed of light: that this is constant regardless of the motion of the observer or the sender. Einstein's second assumption was a restatement of Newton and Galileo's relativity principle: the laws of physics are the same for all observers moving at steady speeds with respect to each other. That is all there is to special relativity: all the surprises and predictions follow from working through the conseqences of these two, seemingly harmless, assumptions. Although it did not appear in his first 1905 paper, Einstein's famous equation relating mass to energy can also be deduced from the two assumptions – with all its ambivalent consequences for humankind.

To show that we need to re-think our ideas of time, consider the following 'thought experiment'. If we lived in a world where the velocity of light was much smaller, 100 miles/hour for example, the effects we are going to describe would be commonplace. There is no loss of predictive power in considering such an evidently impractical thought experiment.

Let us think carefully about what we mean when we say that two things happen at the same time. To check such a statement we must be able to synchronize the two clocks by which we measure the times of the two events. If the two clocks are not in the same place we need to devise a system of signals to check their times. If signals could be transmitted between the clocks instantaneously, there would be no problem. In fact, although light travels very fast, even a light signal takes a small but definite amount of time to travel between the clocks. Since nothing travels faster than light, this is the best we can do. Let us look at what all this means in a simple example (Figure 3.5). We have a very long train with an observer at each end. In the first case, the train is at rest and we station ourselves carefully at the midpoint of the train – one of us on the train and one beside the track. If the signaller at the midpoint of the train sends out a light pulse, it is clear that it will reach both observers at the ends of the train at the same time: the readings on their clocks will be the same as on ours. We can then say that the flash of light arrives simultaneously at each end of the train. Now comes the twist. Suppose the train is moving along at a steady speed. Exactly at the moment the signaller at the midpoint of the train passes the observer by the track, he sends out his pulse of light as before. From our point of view, as an observer by the track, one end of the train is moving towards us and one end is moving away. We therefore find the arrival times to be different since the light pulse has different distances to travel to each end. However, our fellow signaller on the train is

Figure 3.5 (a) The upper picture is of a train carriage at rest, from the centre of which a flash of light is emitted. The lower picture shows that the light pulses reach each end of the train simultaneously.

(a)

(b) The upper picture shows a moving train carriage, from the centre of which a flash of light is emitted. The lower picture depicts the pulse reaching the rear end of the carriage first according to observers stationed on the track. According to those on the train, nothing has changed because of the uniform motion, and so the pulses of light reach both ends of the train simultaneously. (Length contraction of the carriage has been ignored in the figure for clarity.)

(b)

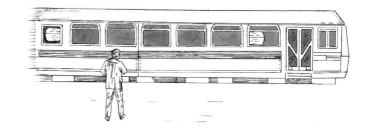

sitting comfortably at rest with respect to the watchers at either end of the train. For him, the signal is received at the same time at both ends, exactly as before. Both are correct! What are simultaneous events for one person need not be simultaneous events for another. This is an inescapable consequence of relativity. It depends on the facts that the speed of light is the same for all uniformly moving observers and that this speed is finite.

A world in which the velocity of light is only 100 miles/hour would seem very strange to us. In such a world, a 'relativistic detective' would have a hard time; for example, taking statements from witnesses about the times of different events would be very confusing. What if a murder had occurred in which someone had been shot? Is it possible that some moving observer could see the victim die before the gun had been fired? Such a violation of 'causality' – cause and effect – is not allowed. We will take up this question again in the next chapter.

Moving clocks run slow

In Chapter 1, we highlighted the fact that relativity predicts that moving clocks run at a slower rate than clocks at rest with respect to us. Let us see how this comes about. Since the effect does not depend on any detail of how clocks work, we shall consider the very simplest form of clock using light beams and mirrors. Again, we will not worry about practical details – if the speed of light were much slower, this experiment would be fine.

Our 'light clock' is shown in Figure 3.6, and consists of two mirrors. We can see directly the consequences of relativity by considering the action of this clock as seen by two different observers. One observer is at rest relative to the clock. He measures one 'tick' of the clock as being the time taken for a light beam to complete the round trip to one mirror and back. How does this look for an observer travelling at some constant speed in a direction at right angles to the line joining the mirrors? She sees the light start out, but then sees both mirrors receding from her (Figure 3.6). According to her, the light has to travel further than twice the distance between the mirrors in order for it to return to its starting point. Since the speed of light is the same for both observers, the time observed between 'ticks' by the moving observer is longer. A simple calculation using Pythagoras's theorem shows how the amount of this 'time dilation' depends on the speed of the observer. This is the famous time dilation result predicted by Einstein's relativity theory. Details of the derivation are given in the Appendix for those interested in

Figure 3.6 A light clock: one unit of time corresponds to a pulse of light travelling from one mirror to the other and back, as shown by the dashed line. When in motion the light path is the longer triangular dashed line shown. Since the light now travels a longer distance when the clock is in motion, the moving clock runs slower than an otherwise identical stationary clock. See the Appendix for more details of the mathematics of this time dilation.

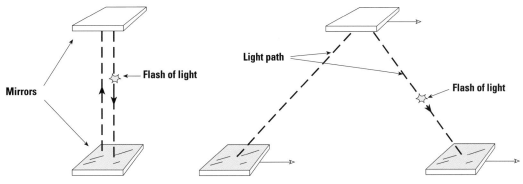

Mirrors / Flash of light / Light path / Flash of light

the mathematics. The Lorentz–Fitzgerald contraction – the fact that the length of a moving object appears shortened – can be deduced in a very similar way.

The physicist George Gamow popularized relativity in his famous book *Mr Tompkins in Wonderland*. The hero of the story is always falling asleep in popular lectures on science by a famous university professor. In one of his dreams, Mr Tompkins is in a world in which the speed of light is reduced to a value for which relativistic effects are a part of everyday life (Figure 3.7). In fact, a detailed study into what one would actually see shows that everything would appear rather differently in Mr Tompkins' world. In such a world, because of the difference in time taken by light rays from different parts of an object, ordinary three-dimensional objects would appear rotated.

As we shall see in the next section, the conventional way in which to view 'events' and draw conclusions about complex situations is through so-called 'space-time' diagrams. Even experienced 'relativists' can still be forced to think hard about the correct explanation of some of the so-called paradoxes of special relativity. At the end of this chapter, we will outline the solution to probably the most famous example of Einstein's theory – the 'twin paradox' mentioned earlier. For now, we conclude this section on time dilation and length contraction with another well-known puzzle.

After our discussion of the aether, we summarized John Bell's contention that concepts such as Lorentz–Fitzgerald contraction and Larmor

Figure 3.7 Two illustrations of length contraction taken from George Gamow's book *Mr Tompkins in Wonderland*, where the speed of light is supposed to be only a little greater than that of the cyclists. The illustration above shows how things look from the point of view of the cyclist. In the picture to the left we see things from a pedestrian's perspective. In fact, things would not look quite as simple with real three-dimensional objects since there are different time delays for light to reach the eye of the observer from different parts of the moving object, as shown in the next illustration. [Courtesy Elfriede Gamow.]

time dilation were the best way to introduce the ideas of relativity. In support of his view, Bell recounted the following anecdote. Theoretical physicists at CERN in Geneva were discussing a well-known puzzle in relativity. The puzzle consists of the flight of two identical spaceships joined by a rope. The two spaceships are at rest, one behind the other, with the rope stretched between them. The spaceships start their engines simultaneously and then both accelerate up to half light speed. According to an observer on the ground, both ships have at every moment the same velocity and so remain separated from one another by the same distance (Figure 3.9) The question to be answered is: 'Does the rope break?' Bell conducted a survey among the scientists in the CERN Theory Division, and there was a clear consensus (but not unanimity) that the rope would *not* break. However, the well-known Nobel Prize winner T. D. Lee, who was visiting CERN at the time, was also asked about this puzzle: after retiring for some time to think about it, he eventually re-emerged to announce that 'either the rope breaks or the spaceships do!' In fact, most physicists, even if they start with a different answer, after playing around with space-time diagrams and considering how the situation looks from the point of view of each pilot, arrive at the accepted answer that the rope breaks. Bell's point is that, from the stationary observer's perspective, the rope undergoes the Lorentz–Fitzgerald contraction, and it is therefore 'obvious' that the rope must break.

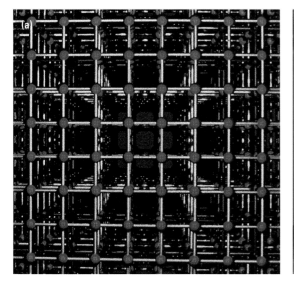 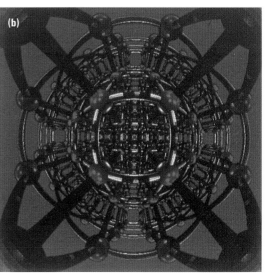

Figure 3.8 The view of a cubic network (a) when travelling at close to the speed of light (b). Relativistic motion seems to induce curious rotations and distortions. Notice also the colour changes. [Photographs by P. K. Hsuing, Pittsburgh Supercomputer Center.]

Space-time

Henceforth space by itself and time by itself are doomed to fade away into mere shadows, and only a kind of union of the two will preserve an independent reality.

Herman Minkowski, 'Space and Time', 1908

Herman Minkowski was one of Einstein's mathematics professors when Einstein was a student in Zurich. He believed Einstein to be a 'lazy dog' since Einstein showed little interest in mathematics and only occasionally bothered to attend his lectures. Minkowski was therefore very surprised by Einstein's success: 'Oh, that Einstein, always cutting lectures – I really would not have believed him capable of it.' As can be seen from the famous (and somewhat over-enthusiastic) quotation at the head of this section, Minkowski took up Einstein's ideas about space and time with a vengeance. It is Minkowski we have to thank for much of the intimidating vocabulary of relativity – for phrases

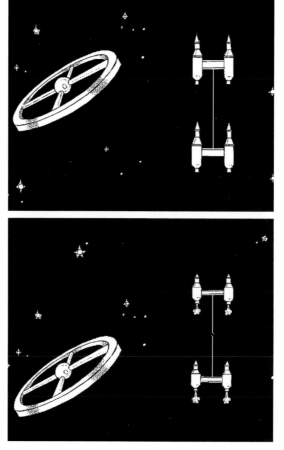

Figure 3.9 Two spaceships are shown tethered together by a taut rope with an observation station nearby. In the lower figure, both spaceships are in rapid motion after having been started simultaneously by the observation station and having followed identical acceleration schedules. According to the observers, both ships remain the same distance apart, although the ships themselves will have undergone a length contraction.

such as 'world lines' and 'four-dimensional space-time continuum'. Minkowski proposed treating time as a variable in the same way as one treats the space 'coordinates' of a point in three-dimensional space: he conjectured that, instead of space and time separately we should now think in terms of 'events in four-dimensional space-time' (see the box on p. 58). Einstein was, at first, unimpressed by such re-writing of his results, and referred to it as 'superfluous learnedness'. He learned to know better: it is arguable that Minkowski's new mathematical formulation of special relativity was a crucial stepping stone for Einstein to go on to solve the problems of general relativity and gravitation.

In some ways, Minkowski was an unlikely popularizer of relativity since he had the reputation of being a very serious and dedicated mathematician. Despite his surprise at Einstein's achievement, Minkowski recognized the truth when he saw it, and immediately immersed himself in studying the further implications of relativity. On one occasion around this time, Minkowski visited an art gallery with fellow mathematician David Hilbert. When Hilbert's wife asked them what they had thought of the pictures, Hilbert replied: 'I do not know; we were so busy discussing relativity that we never really saw the art.' The semi-popular lecture with the famous quotation about space and time was delivered in September of 1908 when Minkowksi was still only forty-four. It is interesting to speculate how things might have developed if Einstein, still in the patent office in Berne, had been able to collaborate with Minkowski in nearby Göttingen. It was not to be. By January of 1909, Minkowski was dead of a burst appendix: legend has it that his last words were 'What a pity that I have to die in the age of relativity's development.'

Einstein was well aware of the bewilderment that the use of Minkowski's new language created. He wrote:

> The non-mathematician is seized by a mysterious shuddering when he hears of 'four-dimensional' things, by a feeling not unlike that awakened by thoughts of the occult. And yet there is no more commonplace statement than that the world in which we live is a four-dimensional space-time continuum.

Figure 3.10 Hermann Minkowski (1864–1909) was born in Lithuania, and, ironically, his family moved to Germany to escape Jewish persecution. Minkowski invented his new approach to relativity when he was a professor of mathematics at the University of Göttingen. He died of appendicitis before his new ideas were published. The famous German mathematician David Hilbert referred to Minkowski as a 'gift from heaven'.
[H.A. Lorentz, A. Einstein, H. Minkowski, Das Relatitatsprinzip, 1915. Courtesy AIP Emilio Segrè Visual Archives.]

Let us see if we can make this apparent. The usual geometry of Euclid concerns points, lines and figures in the three-dimensional space that we are familiar with: we can represent the position of any point by three 'coordinates'. For a hiker, these are just his latitude and longitude, and his height above sea-level. In problems in mathematics and physics, we usually talk about the values of the x, y and z coordinates when specifying positions. What Minkowski proposed was that the position of some happening – an 'event' – should be specified by four coordinates: the three space coordinates x, y and z, describing where the event happened, together with the time, t, of the event, specifying when it happened. Let us get used to this idea of using time as a 'fourth dimension' by looking at the 'world line' of a galloping horse.

A long-standing controversy in the art world of the nineteenth century concerned the motion of horses' legs when galloping. Throughout history, artists had portrayed horses as galloping rather like animated rocking-horses. However, Leland Stanford, an American railway industrialist

Space-time diagrams

The three directions of space (forwards–backwards, left–right and up–down) are very familiar. Mathematically, this idea can be represented by three axes (x, y and z) each at right angles to each other. The position of any object at any given time can be represented by three numbers – the x, y and z coordinates of the object. For a mountaineer, the three numbers (2, 3, 3000) could represent the position of a climber who was 2 miles east, 3 miles north and 3000 feet above the base camp. The whole hike from base to this point can be represented as a curved line in this three-dimensional space. What is missing from this description of the hike is any idea of the time taken by the hiker to reach different points along the path. If we include time as a fourth dimension, each point on the path is now described by four numbers or coordinates. For example, the four-dimensional path could include the 'point' (2 miles E, 3 miles N, 3000 feet, 12 noon August 17th 1996). Such a 'point' that includes the time coordinate is called an *event*. The whole path is called the *world line* of the climber. The world line of those remaining at base camp is just a straight line that follows the time axis. It is difficult to imagine a path in four-dimensional space-time, but if we suppress the height dimension we can draw a perspective view of the path including the time dimension.

   The reader will notice an important difference between the three space dimensions and the time 'dimension'. Hikers can retrace their spatial motion by returning down the mountain on the same path as their ascent. We cannot reverse our motion in the time dimension. Time only advances and we inexorably grow older. Despite this difference, four-dimensional space-time is a very useful concept in relativity.

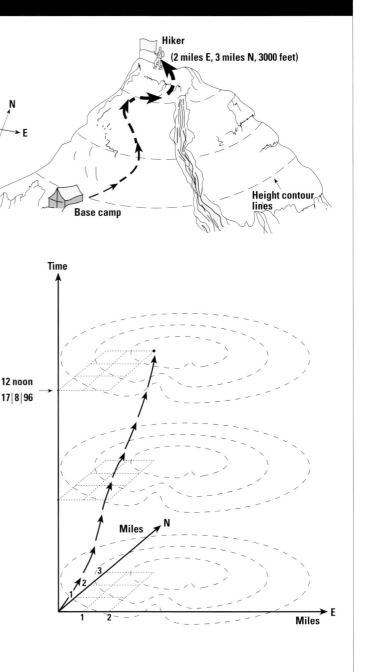

Hiker
(2 miles E, 3 miles N, 3000 feet)

N

E

Height contour lines

Base camp

Time

12 noon
17 | 8 | 96

Miles   N

Miles   E

and founder of the famous University, wished to settle the matter once and for all. In 1872, he decided to use new technology – photography – to do this, and employed a famous photographer, Eadweard Muybridge, to take the photographs. The publication of these photographs in his book *The Horse in Motion* was a genuine revelation to both artists and the public (Figure 3.11). From antiquity, paintings of galloping horses had been wrong: in fact, a horse in fast motion lifts all four legs off the ground at the same time! One side-effect of this discovery was that art academies around the world were deluged by pictures showing horses charging about in every situation, but now with the difference that all four legs were in the air.

But what has this interesting episode in art history to do with relativity and time as the fourth dimension? Look at the sequence of photographs taken by Muybridge. In Figure 3.12, we have arranged the photographs as a space-time diagram, with images higher up the diagram corresponding to later times. Taken together, they represent a 'history' of the horse in time. Notice that we have also displaced each successive image a little to the right, representing the motion of the horse along the racetrack. In technical terms, this is a Minkowski 'space-time' diagram of the horse's motion. Plotting the motion in terms of its position in space and time reveals the 'world line', or life history, of the horse. If you are standing still by the side of the racetrack, your world line on such a plot will just be a vertical line parallel to the time 'axis'. As another, slightly more complicated, example, picture a sequence of images of the Earth moving in its orbit around the Sun. Forming a stack of such images, taken, say, a month apart, builds up the space-time diagram of the Earth's

Figure 3.11 Eadweard Muybridge's famous sequence of photographs of a horse galloping. [Courtesy George Eastman House.]

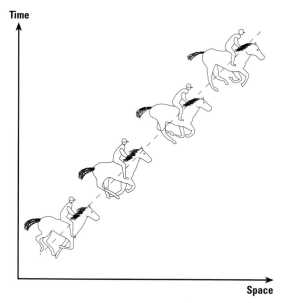

Figure 3.12 The same sequence of photographs of a horse galloping, now arranged as a space-time diagram. Each successive image of the horse has been placed a little further to the right to indicate the distance covered between each photograph.

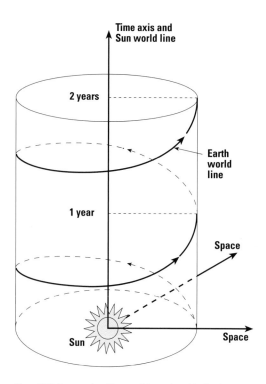

Figure 3.13 A space-time diagram of the motion of the Earth around the Sun.

orbital motion. Since the Sun is at rest in these pictures, its world line is just a vertical straight line. By contrast, since the Earth is circling round the Sun, the Earth's world line traces out a spiral helix (Figure 3.13).

One of the major problems with four dimensions – at least for illustrators – is in visualizing all four dimensions at once. We are all used to the fact that, although a page of paper is two-dimensional, we often draw perspective diagrams showing a third dimension (see the box on p. 58). It is clearly not easy to draw anything that gives us some intuitive idea of a fourth dimension – although mathematicians apparently have no difficulty in considering the mathematics of any number of dimensions. But, as we saw with the spiral world line of the Earth, some idea of a space-time diagram is possible if we need not show all of the three space dimensions. In the case of the Earth, its orbit is (nearly) circular and is restricted to two dimensions – a plane. So, when we add the time dimension, a three-dimensional perspective picture is sufficient to illustrate the trajectory the Earth follows through space and time.

Let us now rejoin the *Starship Enterprise* on a trip to a nearby star. Since the distances between stars are so vast, even a radio signal, which travels at the speed of light would take years to reach us. For this reason, rather than referring to distances as being measured in miles or kilometres – values which would have enormous numbers of noughts – astronomers prefer to measure distances in 'light years', where a light year is defined as the distance light travels in one year. In our more familiar units, 1 light year is equivalent to 9 500 000 000 000 kilometres. So, on our space time plot of the world line of the *Enterprise*, we can choose the *distance* units to be light years and the *time* units to be years (Figure 3.15). In this figure, we have shown the world line of a light signal sent out from Earth at the same time as the *Enterprise* leaves home. With our choice of units, the light signal traces out a line that exactly bisects the angle between the space and time axes: it is a line of unit slope. The world line of the *Enterprise*

Figure 3.14 *Nude Descending a Staircase, No 2* (1912) by the French artist Marcel Duchamp (1887–1968). When this painting was exhibited in New York in 1913 it created a scandal. The picture attempts to represent movement, and this concern to represent the passage of time was a theme in art, as in physics, just before the First World War. Duchamp became associated with Dada and Surrealism, but eventually gave up art to devote himself to chess. [Philadelphia Museum of Art: Louise and Walter Arensberg Collection.]

itself, travelling at half light speed, is also shown. Since the *Enterprise* travels less distance in the same time than a light ray, its world line lies closer to the time direction than that of light. We can now give you a glimpse of the power of the special type of geometric reasoning that such space-time diagrams allow. You should be warned that the next few paragraphs are harder and more technical than the rest of this chapter. If the going becomes too difficult, you are encouraged to turn to the next section.

How does an astronaut on board the *Enterprise* see things? Since she is at rest relative to the spaceship, for her the world line of the *Enterprise* is a straight line, parallel to her time axis. According to this 'Minkowski' view of relativity, we can regard the time direction of the astronaut as being inclined at an angle with respect to the time direction of the observer back on Earth. According to Einstein, both the astronaut and her friend back on Earth will see the light beam moving at the same velocity. For the observer on Earth, the time and space units were carefully chosen so that the light beam travelled along a direction that was exactly symmetric with respect to the time and space axes. This must also be true for the astronaut. Figure 3.16 shows that the time and space axes for the astronaut are in different directions to those of the observer, who remains relatively at rest.

Minkowski's geometric interpretation of relativity is very powerful. We can gain some understanding by looking at a problem from ordinary geometry. Imagine that you are 'orienteering' in the forest – using a map to get from one place to another. Suppose you know where you are on the map, and you have to get to another control point. You can specify how to get there by travelling a certain number of metres in an easterly direction, followed by a different number of metres in a northerly direction. In mathematical language, we call these two distances the 'x and y coordinates' of the point. But orienteering maps can come in two kinds: one variety uses north–south lines aligned according to 'magnetic' north, and the other uses north–south lines parallel to some artificial 'grid' north convenient to the map makers. As can be seen from Figure 3.17, the two sets of x and y coordinates will be different, even though they represent the same physical point on the ground. Obviously the actual distance to the control point must be the same, and this is clear from the figure. By Pythagoras's theorem, the square of this distance is just the sum of the squares of the x distance and the y distance: this is the quantity that remains the same for the two maps. This relationship can be expressed mathematically by the following formula:

$$(\text{distance})^2 = (x\,\text{interval})^2 + (y\,\text{interval})^2$$

In this geometric example, going from one map to another just corresponds to rotating the x and y axes – changing the east and north directions. What Minkowski showed was that the time and space coordinates for observers moving relative to each other could also be thought of as being related by a peculiar type of rotation. It cannot be an ordinary rotation since we have seen that the new time and space directions are no longer at right angles. But, just as for ordinary rotations, for

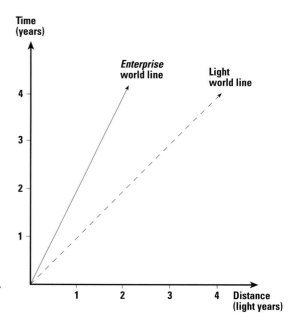

Figure 3.15 A space-time diagram showing the world line for the *Starship Enterprise* travelling at half the speed of light. The world line for a pulse of light which begins its journey at the same time is also shown. Notice that because the distance is measured in light years the light line has a slope of 45 degrees to the vertical time axis. Because no material body can exceed the speed of light, the world lines of any object except a light beam must have a smaller slope with respect to the time axis.

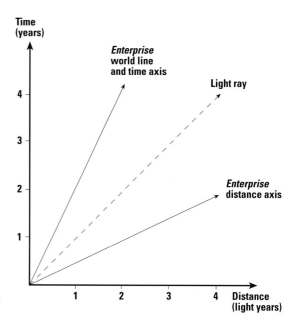

Figure 3.16 For crew members aboard the *Enterprise* the ship is at rest relative to them, and its world line corresponds to their time axis. On this figure, the space axis for those aboard the *Enterprise* is a line such that the light world line is symmetrical with respect to the time line and the space line.

which the distance to a point remains unchanged, so too there is an analogous 'distance' that remains unchanged for these strange space-time 'rotations'. This quantity is said to be an 'invariant', and it is sometimes known as the 'proper time'. Minkowski showed that it is given by the following formula:

$$(\text{proper time})^2 = (\text{time interval})^2 - (\text{space interval})^2$$

It was because of his discovery of this new, absolute, quantity that Minkowski felt that the name 'relativity' was rather a poor name for Einstein's new theory. In the case of ordinary rotations, it is not difficult to see that the new x and y coordinates of a point can be written as a mixture of the old x and y coordinates. In an analogous way, the time and space values of an event seen by a moving observer can be expressed in terms of a mixture of the time and space coordinates of the event as seen by a stationary observer.

It is interesting to conclude this introduction to Minkowski spacetime by considering what happens to Maxwell's electromagnetic fields as seen by different observers. A somewhat similar type of mixing to that of the time and space values occurs for the electric and magnetic fields seen by differently moving observers. Imagine that the *Starship Enterprise* was fired on by an enemy ship. This time the weapon used by the Klingons consists of an electron gun which fires an intense pulse of high-energy, negatively charged electrons.

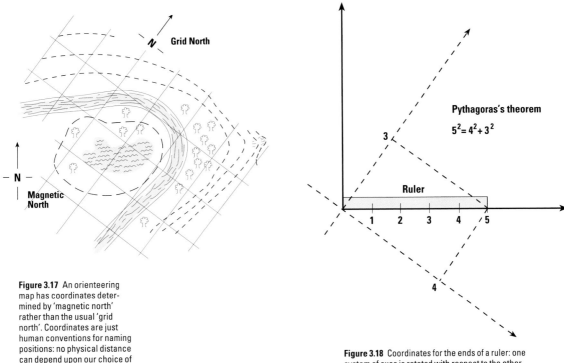

Figure 3.17 An orienteering map has coordinates determined by 'magnetic north' rather than the usual 'grid north'. Coordinates are just human conventions for naming positions: no physical distance can depend upon our choice of coordinates.

Figure 3.18 Coordinates for the ends of a ruler: one system of axes is rotated with respect to the other.

Fortunately, the electron pulse narrowly misses the ship, but the electric and magnetic fields associated with the pulse generate large electric currents that surge about the metallic surface of the *Enterprise* as the electrons stream by in a near miss. Now, according to Einstein, physics should be unaltered if we look at the situation from the point of view of the moving electrons. In this case, we have a stationary cloud of electrons and the *Enterprise* comes rushing past them. We now see a puzzle: as we saw the electron pulse pass by the *Enterprise*, we could measure both an electric field and a magnetic field created by the motion of the charged electrons. But, from the electrons' point of view, all the charges are stationary, so there is only an electric field.

The problem was resolved by Einstein, who showed that electric and magnetic fields are really different aspects of the fundamental electromagnetic field. As Einstein himself said:

> What led me more or less directly to the special theory of relativity was the conviction that the electromotive force acting on a body in motion in a magnetic field was nothing else but an electric field.

The electric and magnetic fields seen by one observer are a mixture of those seen by another observer in relative motion. The precise mixture is actually more complicated to work out than that for the space-time coordinates, but it can be shown that relativity is fully compatible with Maxwell's equations as expressed by both observers. It is a curious thought that, in some sense, all of magnetism can be regarded as a relativistic effect. Relativity can be regarded as a truly everyday phenomenon after all.

The problem twins

> *If we placed a living organism in a box…one could arrange that the organism, after an arbitrary lengthy flight, could be returned to its original spot in a scarcely altered condition while corresponding organisms which had remained in their original positions had long since given way to new generations. For the moving organism the lengthy time of the journey was a mere instant, provided the motion took place with approximately the speed of light.*
> Albert Einstein, 1911

In our first mission on the *Starship Enterprise*, we encountered the famous 'twin paradox', where one twin goes on a journey to a nearby star and returns to Earth to find that the stay-at-home twin has aged faster. This supposed paradox of relativity has probably generated more controversy than any other prediction. Although the result was first stated in Einstein's original paper, this 'twin' or 'clock' paradox became the subject of many papers in the late 1950s. In a collection of seventeen papers on special relativity published in 1963, nine were on this topic alone. It is probably only because the result

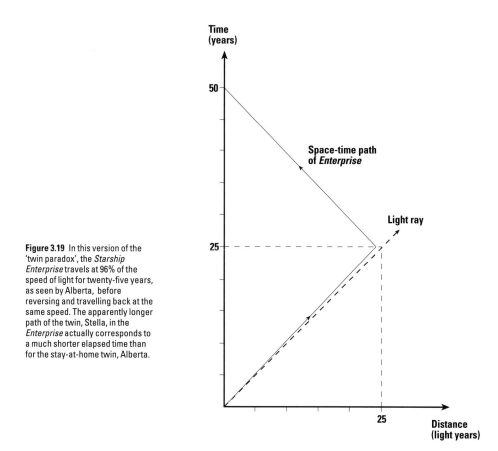

Figure 3.19 In this version of the 'twin paradox', the *Starship Enterprise* travels at 96% of the speed of light for twenty-five years, as seen by Alberta, before reversing and travelling back at the same speed. The apparently longer path of the twin, Stella, in the *Enterprise* actually corresponds to a much shorter elapsed time than for the stay-at-home twin, Alberta.

seems such an affront to commonsense – such differential ageing is so totally outside our everyday experience – that there is such resistance to what is a very straightforward application of relativity. Let us look at this 'paradox' again.

Suppose the twins are called Stella and Alberta. After celebrating their twenty-first birthday together, Stella says goodbye to her sister and climbs aboard the *Starship Enterprise* to begin her journey. The *Enterprise* quickly reaches its constant 'cruising' speed, 24/25 of the velocity of light. Stella then travels for seven years according to her clock on board the spaceship. The spaceship then rapidly slows down, reverses its direction, and travels back to Earth at the same steady speed as before. This again takes seven years of Stella's time, so, when the *Enterprise* returns to Earth, Stella is 21 + 7 + 7 = 35 years old. Figure 3.19 is a space-time picture showing the world lines of both Stella and her sister Alberta, who stayed home. When Stella greets Alberta, she is horrified to find that Alberta is seventy-one years old.

There are many ways to obtain this result. Probably the simplest is to use the time dilation effect. If Stella is travelling at 24/25 of the speed of light then each year of her time takes 25/7 years for Alberta. (This may be derived by directly substituting $v = (24/25)c$ in the time dilation formula given in the Appendix.) Alternatively, we could use length contraction to arrive at the same result. For the traveller, Stella, the distance travelled is only $7/25$ the distance calculated by Alberta. Either way, Stella ages only fourteen years compared with Alberta's fifty.

A question often asked is why we cannot use relativity to think of Alberta as the traveller and Stella at rest in the spaceship? The answer is that the twins are not equivalent. At the turnaround point, Stella can feel a change in the motion of her spaceship: nothing happens to Alberta. There is no paradox – the twins have different experiences. This can also be seen from the space-time picture of Figure 3.19. It is clear that the paths of Stella and Alberta through space-time are different. If this diagram represented the routes taken by Stella and Alberta travelling by car two different ways from one town to another, it would come as no surprise that they took a different time. The surprise in relativity is that the apparently longer path in space-time taken by Stella takes less time than Alberta's. This is due to the fact that Minkowski's special geometry must be used to calculate distances in space-time. Roughly speaking, instead of using the Pythagoras theorem with its sum of two squares, we must use the difference of two squares to calculate the time difference seen by Stella and Alberta.

There are many more puzzles and 'paradoxes' in relativity. One example is the so-called 'pole–barn paradox'. A runner carrying a 20 metre pole runs so fast that the pole appears only 10 metres long to a stationary observer. At some instant, therefore, it seems that we could enclose the pole in a barn that is only 10 metres long. But consider what the runner sees. He sees the pole as 20 metres long and the barn contracted to 5 metres. How can a 20 metre pole fit in a 5 metre barn? In this case the problem is our intuitive notion of simultaneity – which is inadequate for speeds near the velocity of light. To resolve this puzzle requires a careful space-time analysis that keeps track of the times at which the front and back of the pole enter the barn. There are many similar problems to start people thinking!

In a world where the velocity of light was very much reduced, we would be forced to confront these surprising tricks of Nature on a day-to-day basis. In Chapter 1, we discussed some of the experimental evidence that confirms these startling predictions of Einstein's theory. It was only because these relativistic effects are so small, and because Newtonian absolute time is such a good approximation, that the true nature of space-time remained hidden from us for so long. So what was Einstein's great contribution? The British physicist Sir Edmund Whittaker, in a biographical memoir for the Royal Society, wrote that Einstein had 'adopted Poincaré's Principle of Relativity as a new basis of physics'. This view – that Einstein merely reinterpreted the work of Lorentz and Poincaré – was not held by many. More typical was the reaction of Max Born, famous for his probability interpretation of quantum mechanics. Born recalls that:

> Although I was quite familiar with the relativistic idea and the Lorentz transformations, Einstein's reasoning was a revelation to me. The exciting feature of this paper was not so much its simplicity and completeness, but the audacity of challenging Isaac Newton's established philosophy, the traditional concepts of space and time. That distinguishes Einstein's work from his predecessors and gives us the right to speak of Einstein's theory of relativity.

4   The ultimate speed

After ten years of reflection such a principle resulted
from a paradox upon which I had already hit at the age
of sixteen: If I pursue a beam of light with the velocity c
[the velocity of light in a vacuum], I should observe such
a beam of light as a spatially oscillatory electromagnetic
field at rest. However, there seems to be no such thing,
whether on the basis of experience or according to
Maxwell's equations.

Albert Einstein, *Autobiographical notes*, 1949

The strange behaviour of the velocity of light

As we have seen, Roemer showed as long ago as 1676 that the velocity of light
was not infinite. Subsequent measurements by Michelson and others now
agree on a value for the speed of light of some 299 792 kilometres/second.
This applies not only to the visible part of the electromagnetic spectrum but
also to much longer-wavelength radio waves and much shorter-wavelength
gamma rays, as expected from Maxwell's equations. Now, according to
Newton's laws of motion, there is nothing special about the speed of light.
There is nothing, in principle, to stop one accelerating an object – or indeed
oneself – to any speed whatsoever. It was the problem of what one would see
in a mirror if both observer and mirror were moving at the speed of light that
set Einstein on his path to relativity.

It is sometimes said that Einstein showed little exceptional talent
when he was at school. This may be true, but it is certain that few schoolboys
could have formulated the key paradox of the mirror at the age of sixteen.
The paradox related to the idea that the velocity of light was fixed relative to
the aether, just as the speed of sound is fixed relative to the air. In the case of
sound, a jet aircraft travelling at Mach 1 – the speed of sound – keeps pace
with its own sonic bang. By analogy, light from the observer would never
reach the mirror, and the observer would be unable to see an image. The
same thing is said, in slightly more technical terms, in the quotation at the
beginning of this chapter. Einstein did not believe that such new phenomena
could be brought into existence simply by uniform motion. His solution was

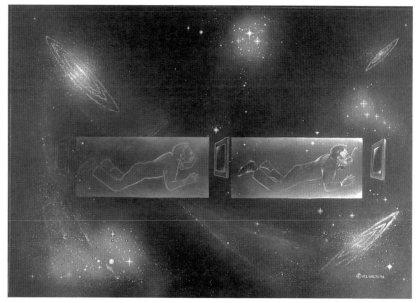

Figure 4.1 The schoolboy Einstein's famous thought experiment with a mirror: if he moves through the aether at the speed of light, what will he see in the mirror? The fainter Einstein image on the left shows the situation a few moments prior to the right-hand image and illustrates that Einstein, the pulse of light and the mirror all move in tandem at the speed of light. [Painting by Marie Walters.]

Figure 4.2 David Mermin wrote the memoir about the great Russian physicist, Landau, referred to in Chapter 1. When asked for a short biography by the editor of the journal *Physics Today*, he provided the following: 'N. D. Mermin is professor of physics at Cornell University. Struggling to qualify as a 4½ he has made occasional contributions to the theory of phase transitions and liquid helium.' In fact, Mermin is a very able and thoughtful physicist with many significant research achievements to his credit. In addition he has a talent for communicating something of the excitement of modern physics to both physicists and non-physicists. His battle to have the word 'boojum' accepted in the distinguished pages of *Physical Review Letters* – the most authoritative and widely circulated of all the international physics journals – is legendary. Mermin has also written several thought-provoking and illuminating articles on aspects of both quantum mechanics and special relativity. [Photograph by David Lynch-Benjamin.]

both radical and elegant: he proposed that the velocity of light was a universal constant, independent of the aether and of the motion of either the light source or the observer. The null result of the Michelson and Morley Experiment is now easily explained: the velocity of light is exactly the same for both light paths so there is no shift of the interference fringes. Now, as we have seen in the preceding chapter, Einstein's proposal leads to time dilation. However, if we try to insert the speed of light into the formula $1/\sqrt{[1 - (v^2/c^2)]}$, the time dilation factor becomes infinite – in a sense, time stands still! This is a strong hint that, as we shall see later, it is impossible to accelerate any massive object, such as ourselves, to light speed. The resolution of the mirror paradox is that at any speed close to, but less than, the speed of light, the observer sees the same image as he sees at rest, and he is never able to reach the speed of light.

What are the implications of Einstein's suggested behaviour for light? In the preceding chapter, we saw that Einstein's theory leads directly to both time dilation and length contraction. We now explore another of the startling consequences of this assumption: a modification of Newton's addition law for velocities. Not only is the velocity addition law modified, but also the speed of light appears as the ultimate speed – a speed that cannot be exceeded. There are many ways to arrive at Einstein's result. One approach relies on a direct application of time dilation and length contraction; another uses the so-called 'Lorentz transformations' that relate the coordinates of events for moving observers. The approach that shows most directly the relationship between the constancy of the velocity of light and Einstein's velocity addition formula is due to David Mermin. If our argument becomes too hard-going, you are advised to turn to the next section.

Mermin's argument uses a gedanken experiment similar to that used in our discussion of simultaneity. We consider a long train moving at speed v relative to a stationary observer (see Figure 4.3). At a certain time, a photon – a particle of light – with speed c and a massive particle with speed w begin a race from the rear of the train to the front. The photon, moving at the speed of light, wins the race and is reflected back towards the rear of the train. It encounters the massive particle moving towards the front of the train when it is still a fraction f of the length of the train from the front. There can be no dispute by any observer where the two particles meet: all must agree on the fraction f. This fraction can be calculated in terms of v, w and c from three simple facts:

1. The total distance that the massive particle moved from the start of the race to its encounter with the photon is equal to the distance that the photon moved in going from the rear to the front, minus the distance that the photon moved in going from the front, back to the particle.

2. The distance that the photon moved in going from the rear to the front is just the length of the train plus the distance that the train moved while the photon was travelling.

3. The distance that the photon moved in going from the front to the massive particle is just the length of the train from the front to the meeting point, minus the distance that the train moved while the photon was so moving.

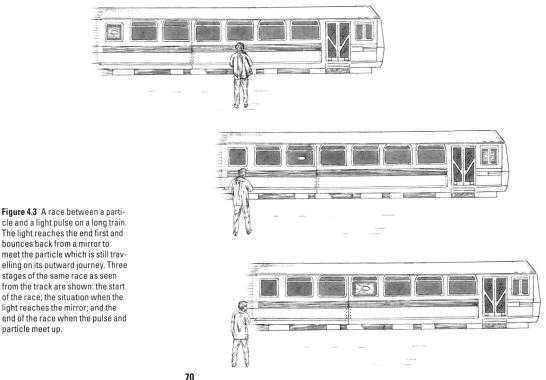

Figure 4.3 A race between a particle and a light pulse on a long train. The light reaches the end first and bounces back from a mirror to meet the particle which is still travelling on its outward journey. Three stages of the same race as seen from the track are shown: the start of the race; the situation when the light reaches the mirror; and the end of the race when the pulse and particle meet up.

70

Calculating this fraction both from the point of view of the stationary observer and from the point of view of an observer on the train gives the result we want. Instead of the Newtonian answer

$$w = u + v$$

where u is the speed of the massive particle as seen on the train, we obtain the following result:

$$w = (u + v)/(1 + uv/c^2)$$

The mathematics is not difficult to follow: it is included in the Appendix for interested readers.

We can now understand the strange results claimed in our imagined battle between the *Enterprise* and the Klingon starship in chapter 1. Even if one of the speeds u or v is equal to the velocity of light, c, the observed velocity w never exceeds c. If both v and u are equal to c, the observed velocity is not $2c$, but just c: this is Einstein's mirror replayed in mathematics. We are forced to the conclusion that the velocity of light cannot be exceeded, and that Newtonian mechanics must be wrong. As we shall see in the next chapter, following through the logical implications of this conclusion leads inexorably to Einstein's famous result connecting mass and energy. Before we take this up, we must look at what experimental evidence supports Einstein's conjecture about the speed of light.

Binary stars and the neutral pion

Figure 4.4 The Dutch astronomer, Willem de Sitter (1872–1935), with his wife. De Sitter's main scientific passion was a study of the satellites of Jupiter, but he is now remembered mostly for his work on general relativity, including a model for an empty expanding universe. [Courtesy Sky Publishing Corp.]

Does light behave in the way Einstein suggested? In 1913, eight years after Einstein's paper, the Dutch astronomer Willem de Sitter found evidence that the speed of light is indeed independent of the speed of the source. His experimental equipment consisted of 'binary' stars – double star systems whose stars revolve about each other like an Earth–Moon system. The first double star was discovered in 1650 by the Italian astronomer Jean Baptiste Riccioli. By 1821, the British astronomer William Herschel had published catalogues listing more than 800 such binary stars. De Sitter pointed out that some of these pairs of stars have orbits aligned so that each star is either coming towards us or going away from us as it goes round its orbit. If the speed of light is modified by the motion of the star then 'faster' light from positions when the star is approaching the Earth could catch up with 'slower' light emitted earlier when the star was moving away from us in its orbit (see Figure 4.5). There would then arise the possibility of multiple 'ghost' images of the star in different positions. Such ghost images have not been observed: binary stars always appear to be moving in well-behaved orbits about each other. Because of this, de Sitter concluded that Einstein was right in his assumption about the speed of light.

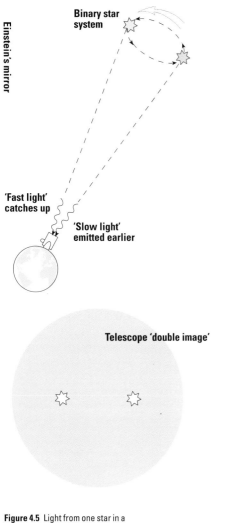

Binary star system

'Fast light' catches up

'Slow light' emitted earlier

Telescope 'double image'

Figure 4.5 Light from one star in a binary system arrives at a telescope on Earth. (The other star in the system is not shown.) If the speed of light depended on the speed of the star, then light from times when the star is receding from Earth could be overtaken by faster light emitted later, when the star was approaching the Earth. In our telescope on Earth the picture of the binary system would be confused by 'ghost' images.

In fact, it has since been argued that one needs to be careful in accepting de Sitter's binary star argument as evidence supporting Einstein. This is because of a phenomenon called 'extinction'. Light travelling through matter undergoes a continual process of absorption and re-emission by the molecules of the matter. Not only does the velocity of light appear to be reduced from its vacuum value by this process as it travels through the medium, but, more importantly in this case, only a small thickness of matter is sufficient to get rid of any 'memory' of the motion of the original source. Since binary star systems are usually surrounded by gas clouds, these clouds may remove any trace of an effect due to the motion of the binary star. Thus de Sitter's famous 'proof' of Einstein's postulate has been called into question. To avoid this problem, the binary star experiment has recently been repeated with stars emitting X-rays, which are expected to be almost unaffected by the interstellar gas. An experiment performed in 1977 conclusively confirmed Einstein's postulate.

The most direct test of Einstein's proposal for the universality of the velocity of light comes from experiments in elementary particle physics. One of the particles responsible for the nuclear force is called the pion. This is a particle that is considerably more massive than an electron and is about one-seventh of the mass of the proton. Protons are believed to be made up of three quarks – fractionally charged building blocks for elementary particles. The pion is an example of a so-called 'meson', and is believed to contain a quark and an anti-quark. Pions come in three charge states, +1, −1 and 0, in units of the proton's electric charge. We are interested in the neutral pion. This is a very short-lived elementary particle (see Figure 4.7) which 'decays' into two photons – two particles of light. If we can measure the speed of the photons resulting from the decay of a pion in flight, we can test Einstein's hypothesis. This experiment has been performed, and neutral pions travelling at more than 99% of the speed of light were created by high-energy protons, produced by an accelerator, colliding with a stationary target. The speed of the photons from the decay of the pions was carefully measured by timing the photons over a flight path

Figure 4.6 The yellow-orange star (Algol B) eclipses its more brilliant blue companion (Algol A). The more distant white star (Algol C) is the third star in this triple system. [Painting by Marie Walters.]

in air of about 30 metres. For such energetic photons the 'extinction length' – the length at which they would have lost all memory of the motion of their source – is some 5 kilometres. There is therefore no ambiguity about the interpretation of the result of the experiment. The relativistic velocity addition law is found to be verified to very high accuracy; and velocity addition according to Newton and Galileo is proved to be inadequate at these high speeds.

Doppler and Einstein

There are still more ingenious tests of Einstein's proposal about the velocity of light. These involve a phenomenon called the Doppler effect. In 1845, in Utrecht, a troupe of trumpeters played their instruments while being transported by a train. Another group of musicians arrayed alongside the railway track were paid to listen carefully as the train and the trumpeters went by. This seemingly bizarre experiment was the invention of the Austrian physicist Johann Christian Doppler. Most readers will have noticed the drop in pitch of a whistle as a train goes by, or of a siren as an ambulance or police car passes. It was precisely this same change in pitch with motion that Doppler was investigating with his entertaining train and trumpet experiment.

Sound is a pressure wave that must travel through some medium such as air. When the object producing the sound is approaching us, the sound waves are bunched up closer than normal because of the motion of the source. As a result, the effective wavelength λ is shorter and the frequency f that we hear is higher than if the object had been at rest. This is because of the well-known relationship between wavelength and frequency for a wave travelling with velocity v

$$v = f\lambda$$

Similarly, if the object producing the sound is moving away from us, the waves are more spread out and we hear a lower frequency. This is the Doppler effect. Now imagine that the sound is produced by an aircraft moving at exactly the speed of sound. The waves will all be compressed together in the forward direction, and all the waves will arrive where we are standing at the same time, causing a sonic boom. As the plane recedes away from us, we receive Doppler-shifted sound waves, in spite of the fact that the aircraft is moving away with the speed of sound. We would continue to hear the sound even if the aircraft moved away faster than the speed of sound. On the other hand, if the aircraft were to drop some bombs and then fly away at a speed faster than the speed of sound, the sound waves from the explosion would never catch up with the aircraft. We see that for sound there is a lack of symmetry between the effects of motion of the source and motion of the observer. The reason for this was apparent in our thought experiments in chapter 2.

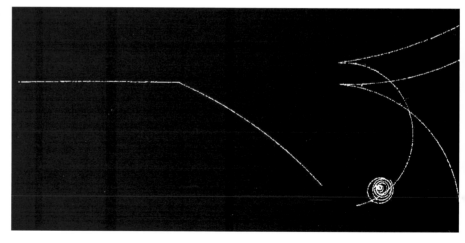

Figure 4.7 The tracks in a bubble chamber picture are caused by charged particles agitating the atoms as they pass; this heating causes bubbles to form preferentially along the tracks of the atoms when the pressure on the chamber is released. Neutral particles leave no tracks, and their presence can only be deduced indirectly. In this picture a neutral pi meson is produced in the decay products of another type of meson called a kaon (K). The diagram to the right of the bubble chamber photograph (see p. 75) uses energy and momentum conservation to reconstruct the decay process. The neutral pion (π°) then decays almost immediately into two photons (γ). The photons are neutral and also leave no tracks in the chamber: in this case, their presence is revealed by their interaction with a lead sheet in the chamber. The photons are converted to electron–positron pairs, and the negatively charged electrons and positively charged positrons (anti-electrons) curve in opposite directions in the chamber's magnetic field. [Ernest Orlando Lawrence Berkeley National Laboratory.]

Sound needs a medium such as air in which the waves can propagate and thus the speed of sound is fixed with respect to the stationary air.

Doppler believed that light would also experience a Doppler effect. This was natural since in the nineteenth century, light was thought of as being closely analogous to sound, i.e. a wave motion with the aether playing the role of the medium. In fact, as we now know, things are not so simple. Not only does light show both wave and particle aspects in its behaviour, according to quantum mechanics, but Einstein also abolished the aether. There *is* a Doppler effect predicted for light, but, unlike its counterpart for sound, its effect is perfectly symmetric with respect to the motion of the source and the receiver. This follows directly from Einstein's relativity: the speed of light is unaltered by the motion of the source – as for sound – but also, in the case of light, unaltered by any motion of the receiver.

The aforementioned experiment with the X-ray binary star can be improved if one makes a careful study of the frequency of the light received from the double star system. As the star moves round its orbit, a Doppler shift in observed frequency is expected according to the usual arguments. However, if the velocity of light depends on the motion of the source, a different periodic Doppler shift in frequency is predicted as the star goes round its orbit to that anticipated by Einstein. Experiment confirms Einstein's prediction.

Relativity also gives rise to another, previously unsuspected, Doppler effect. This arises from time dilation effects due to motion of the source. Suppose the *Starship Enterprise* were travelling across the line of sight from Earth. Since the starship is neither approaching nor moving away from the Earth, no 'normal' Doppler effect is anticipated. But imagine that the *Enterprise* sends regular clock signals to the Earth. Because of time dilation,

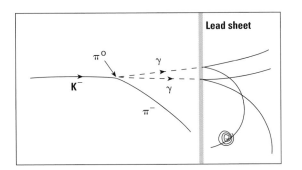

Figure 4.8 Christian Johann Doppler (1803–1853) was an Austrian physicist who was frustrated at only being able to get junior teaching posts. He was on the point of emigrating to America when he was offered a senior teaching post at a school in Prague. He eventually became Professor of Experimental Physics in Vienna. His discovery of what is now known as the Doppler effect was originally for sound waves. Applied to light, the Doppler effect has led to many significant discoveries in astronomy. The Doppler effect is now a routine feature of many modern technological devices, such as police radar speed traps. [Courtesy AIP Emilio Segrè Visual Archives.]

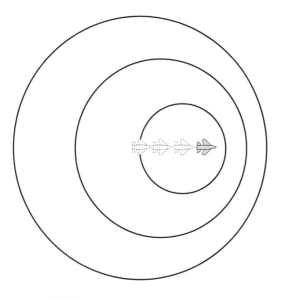

Figure 4.9 The Doppler effect for sound from an aircraft travelling at half the speed of sound – Mach number ½. The dotted images of the aircraft indicate its earlier positions, from which circular (spherical) wavecrests emerge. The diagram shows that the waves are compressed ahead of the plane and stretched out along the trail of the aircraft.

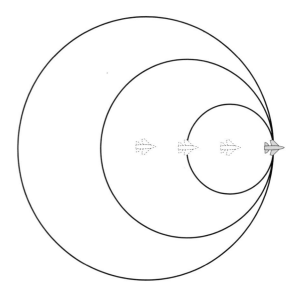

Figure 4.10 Sound waves expanding from a jet travelling at Mach 1 – the speed of sound. The dotted jet images mark the successive positions of the jet as each wavecrest is emitted. The waves pile up at the front of the jet and are stretched out in its wake.

these clock signals arrive at the Earth at longer time intervals than expected: for the observer on Earth they are shifted to lower frequency, or 'red-shifted'. This is called the 'transverse' Doppler effect. Time dilation also causes a modification to the ordinary Doppler effect. Another contrast to the Doppler effect for sound occurs if the source and the receiver are separating at the speed of light: in this case, the red-shift is predicted to be infinite.

A remarkable example of the transverse Doppler effect occurs in a star system known as SS433 (the 433rd entry in a catalogue of stars with unusual spectra compiled by the American astronomers Bruce Stephenson and Nicholas Sandulaek). In 1978, two British astronomers, Paul Murdin and David Clark, identified SS433 with the position of an X-ray source embedded in the remains of a 'supernova' explosion. A supernova is the most violent of all star explosions and is believed to leave behind a compressed compact ball of hot neutrons – a neutron star. Murdin and Clark observed unusual behaviour in the spectral lines of SS433 but were unable to continue their observations long enough to explain the phenomena in detail (see the box on p. 78). An American, Bruce Margon, then showed that SS433 had emission lines, including those of hydrogen, that occurred in pairs – one set shifted towards the red end of the spectrum and the other towards the blue. Moreover, these pairs of emission lines shifted cyclically, over a period of 164 days, about a median point red-shifted from the 'rest' wavelength observed in a terrestrial laboratory.

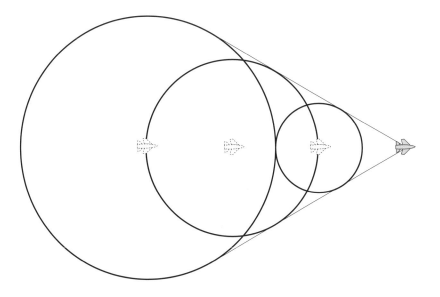
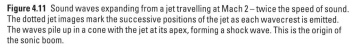

Figure 4.11 Sound waves expanding from a jet travelling at Mach 2 – twice the speed of sound. The dotted jet images mark the successive positions of the jet as each wavecrest is emitted. The waves pile up in a cone with the jet at its apex, forming a shock wave. This is the origin of the sonic boom.

Figure 4.12 Star system SS433. Matter flows from the giant star on to an accretion disc surrounding a neutron star. Narrow jets shoot out of the disc at a relativistic speed of about one-quarter that of light. The approaching jet is blue-shifted and the receding jet correspondingly red-shifted. [Painting by Marie Walters.]

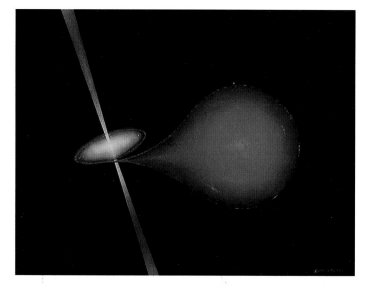

Spectra and stars

The story of the discovery of spectra – the characteristic colours of light emitted by different elements – begins with some experiments conducted by the German physicist Gustav Kirchhoff and his chemist friend Robert Bunsen – of 'Bunsen burner' fame. They discovered that, when a substance is burnt in a flame, the atoms that comprise the substance emit light with characteristic colours. They identified the different colours by passing the light produced through a prism to reveal the various different wavelengths of light that were present, producing a characteristic pattern of lines. Each element produces a different pattern of lines, and this is known as the 'emission spectrum' of the element. Using these spectra to identify which elements were present is similar to taking fingerprints to find out who was at the scene of a crime. A familiar example of the characteristic spectrum of light emitted is the yellow light of heated sodium, often used in street lighting.

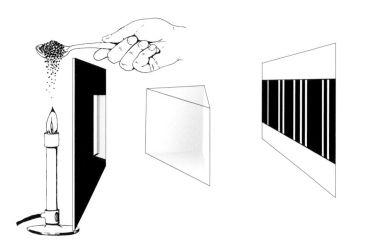

We can also measure a sort of inverse of emission spectra. If white light is passed first through a flame of burning material and then through a prism, the resulting 'rainbow' of colours is interspersed with dark lines. These dark lines occur at exactly the same places in the spectrum as the emission lines for the heated material. They are the result of the light of those particular wavelengths being absorbed from the original white light. These inverse spectra are called 'absorption spectra', and they can also be used to identify the presence of different elements in a substance. The element helium was discovered in the absorption spectrum of light from the Sun.

The use of emission and absorption spectra to study the stars was pioneered by a wealthy London amateur astronomer called William Huggins. He had described his feelings on hearing of Kirchhoff's discoveries about spectra as being 'like coming upon a spring of water in a dry and thirsty land'. In 1868, Huggins observed that the patterns of the spectral lines of light from the Dog Star, Sirius, clearly corresponded to the patterns of spectral lines produced by the same elements found in his laboratory on Earth – with one important difference. The wavelengths of all the lines from Sirius were shifted towards the red end of the spectrum. Huggins realized that this must be due to the Doppler effect; i.e. that Sirius must be moving away from the Earth. This episode marked the birth of the now routine techniques used by astronomers to measure the speeds of stars and galaxies. And it was by looking at light from many different galaxies that the American astronomer Edwin Hubble arrived at his suggestion that the universe was expanding.

Sir William Huggins (1824–1910) pioneered the study of the stars from their spectra. Spectroscopy also led to romance for Huggins. He met his future wife, an enthusiastic amateur astronomer, on a trip to Dublin. She had just started work on spectroscopy, guided by an article written by Huggins. Working as a team, they were pioneers of the new subject of astrophysics. [Courtesy The Mary Lea Shane Archives of the Lick Observatory.]

An explanation for the puzzle of SS433 was suggested by two Cambridge astrophysicists, Andy Fabian and Martin Rees, and independently by Mordechai Milgrom of Israel. SS433 is a binary star system with one star being the dense remnant of a supernova explosion – a neutron star. Matter flows towards the neutron star from its companion, and this results in two symmetrical jets appearing from the neutron star, although the exact mechanism for how these jets are produced is not well understood. Tidal forces due to the gravitational attraction of the companion star cause the two jets to rotate with a period of 164 days. This explains the paired red- and blue-shifted components of the spectra. But why does the spectrum show a basic redshift? The answer is that the material in the jets is travelling so fast that, even when there is no component of the motion in the direction of the Earth, a transverse Doppler shift occurs due to time dilation. Radio telescope images of SS433 (see Figure 4.13) show blobs of matter shooting out from a central source with a speed of about one-quarter that of light. Understanding such complex star systems is rather like solving a real-life detective story in which many different, and perhaps unrelated, things are happening at the same time.

Figure 4.13 Four radio pictures of SS433. The jets extend about one-sixth of a light year from the central source. [Courtesy NRAO/AUI. Observations by R. M. Hjellming and K. J. Johnston.]

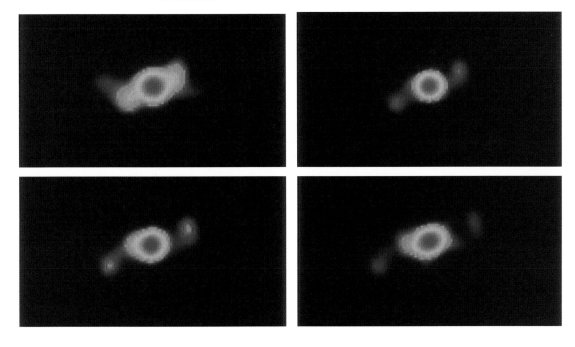

Faster than light

He had steeled himself for the Jump through hyper-space, a phenomenon one did not experience in simple interplanetary trips. The Jump remained, and would probably remain for ever, the only practical method of travelling between the stars. Travel through ordinary space could proceed at no rate more rapid than that of ordinary light (a bit of scientific knowledge that belonged among the few items known since the forgotten dawn of human history), and that would have meant years of travel between even the nearest of inhabited systems. Through hyper-space, that unimaginable region that was neither space nor time, matter nor energy, something nor nothing, one could traverse the length of the Galaxy in the intervals between two neighbouring instants of time.

Isaac Asimov, *Foundation*

Any *Star Trek* fan will be aware that faster-than-light jumps through 'hyper-space' are a routine fact of life aboard the *Enterprise*. Such a device is, of course, absolutely necessary for much of contemporary science fiction, given the distances and sub-light-speed travelling times between Earth and even the nearest stars. But is such faster-than-light travel anything more than science fiction? In certain circumstances – although admittedly none of any use to science fiction writers – we shall see that faster-than-light speeds are indeed possible. Relativity and science fiction will be explored in more detail in chapter 11.

We shall begin by raising an apparently unrelated question: what causes the eerie blue glow of a uranium nuclear reactor core cooled by heavy water? In the early 1930s, many scientists had observed a similar blue light when ordinary water was exposed to radioactive materials. In 1934, a Russian physicist, Pavel Cherenkov, realized that the blue glow was caused by charged particles travelling through the water at a speed greater than that of light when travelling through water (see Figure 4.14). To understand how this comes about, we need to look in more detail at what happens as light travels through matter.

We have already mentioned that, when light travels through matter, it is continually being absorbed and re-emitted. The electric and magnetic

Figure 4.14 The blue glow from this fuel storage pond is due to Cherenkov radiation. [AEA Technology.]

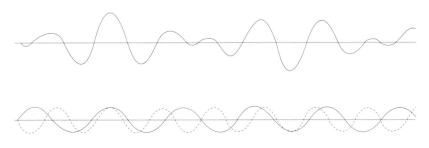

Figure 4.15 The bottom diagram shows two pure waves. The top diagram shows the wave resulting from the sum of the two waves below. Notice that a peak in this sum occurs whenever the two component waves vibrate in step. To form a single pulse, a very large number of pure waves with different wavelengths must be added together. The resulting pulse moves with a speed – called the group velocity – that is different from that of the pure component waves.

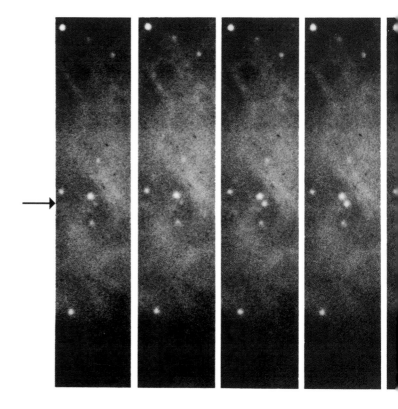

fields of the light wave interact with the electrons of the molecules of the
material through which it is passing, causing the electrons to 'vibrate'. The
accelerated charges – the vibrating electrons – create new light waves, and
these re-radiated waves interfere with the original wave, resulting in both a
change in the velocity of the light wave and a net absorption of the wave energy. Transparent materials, such as water or glass, absorb light much less than
materials which are opaque to visible light: in metals, for example, visible
light can only penetrate a tiny distance. The *apparent* velocity of visible light
through transparent material is found to be less than the speed of light in free
space. We have used the phrase 'apparent velocity of light' since the resulting
wave motion in the material arises from the interference of two component
waves that are both travelling with the usual speed of light. The aforementioned interference effect is found to depend on the frequency of the light.
If we replace visible light by X-rays, we find that many materials that are
opaque to visible light, such as carbon in the form of graphite, are transparent to X-rays. More surprisingly, we find that the effective speed of X-rays in
the substance turns out to be greater than the speed of light in free space!
Does this mean we can send signals at speeds faster than light? The answer
turns out to be no. The reason is that to send information in a signal we need
more than just a wave of one frequency. In order to send a signal containing
information we need to change a uniform simple wave-form into something

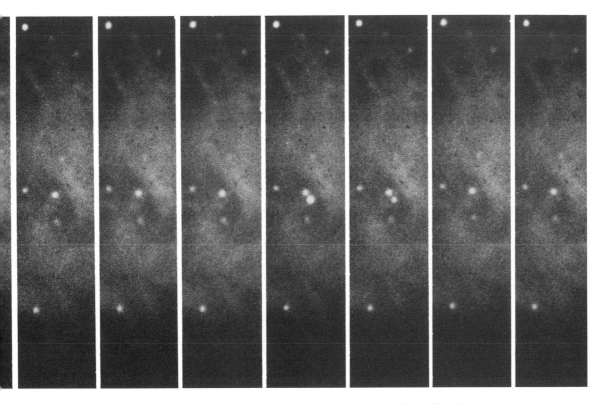

Figure 4.16 This sequence of optical images of the Crab nebula shows the periodic flashes of the pulsar over a whole cycle. The Crab pulsar flashes about thirty times a second across the whole electromagnetic spectrum from radio waves to X-rays. [NOAO photo.]

Figure 4.17 The diagram shows a rotating pulsar, together with the speed-of-light cylinder. In order to make the diagram clear, we have shown only one magnetic field line emerging from the poles, which we have also chosen to be perpendicular to the rotation axis. The magnetic field lines will curve onto the speed-of-light cylinder. One model of pulsars suggests that the radiation emerges near the speed-of-light cylinder. We have shown a narrow beam of radiation occurring at only two points in order not to clutter the illustration.

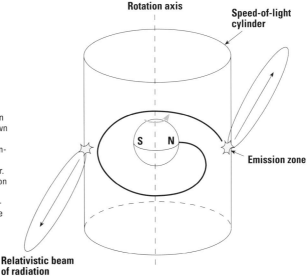

◀ **Figure 4.18** An artist's impression of the central region of a quasar showing two jets emerging from the core. Quasars are brighter than one hundred galaxies, but their enormous power is emitted from a volume not much larger than the solar system. [Painting by Michael Carroll.]

that is more 'bunched up' so we can send pulses of information with definite start and stop times. To make such pulsed wave-forms requires a combination of waves of different frequencies. We can work out in detail what happens as this pulse travels along by adding up all the effects of the different waves making up this pulse (Figure 4.15). The speed of the pulse – called the 'group velocity' – turns out not to be the same as the average speed of the waves making up the pulse! This group velocity – the speed at which information can be transmitted – is never greater than the speed of light in a vacuum.

So, what is happening in the case of radioactive particles and water? The electrons emitted from the 'decays' of radioactive matter are moving faster than the apparent speed of light in the water. This causes a shock wave in the same way that a jet aircraft causes a sonic boom when it is moving faster than the speed of sound. This phenomenon is called the 'Cherenkov effect' after its discoverer, and the effect is now routinely used by particle physicists to determine the velocity of elementary particles in their experiments.

An even simpler example of faster-than-light travel occurs when a lighthouse beam sweeps across a very distant screen. This is not as fanciful as it sounds: Nature has provided us with such stellar lighthouse beams in the form of pulsars. A pulsar is believed to be a rotating neutron star from which a narrow beam of radiation emerges. The pulsar in the Crab nebula is roughly 6500 light years away from us, and its beam sweeps across the Earth about thirty times a second. As the beam crosses the Earth, it is clearly moving much faster than light. However the beam is made up of many millions of photons – particles of light – all of which have been travelling for some 6500 years. Because of the large distance of the pulsar from Earth, the beam appears to sweep across the Earth at faster-than-light speed yet the individual photons are all moving at light speed from the pulsar to the Earth. No single physical object moves across the Earth at faster-than-light speed. However, pulsars do exhibit interesting relativistic effects. At the distance from the star where the motion of the sweeping photon beam reaches the speed-of-light, relativity comes into play. For the Crab pulsar this speed of light distance corresponds to a distance 1500 kilometres from the centre of the neutron star. Intense magnetic fields generated by the neutron star must rotate with the star and curve onto this velocity-of-light 'cylinder', as shown in Figure 4.17. Beyond this 'cylinder', the magnetic field lines must somehow shear and break. Although the question of the location of the generation of the pulsar radiation is still disputed, many astrophysicists believe that this velocity-of-light cylinder is involved.

In about 1970, several groups of radio astronomers discovered radio sources which seemed to be separating at a speed faster than that of light. These so-called 'superluminal' sources are associated with strange objects known as quasars. Quasars, discovered in 1962 by Maarten Schmidt, are enigmatic sources which appear to emit more power from a region about the size of the solar system than does an entire galaxy. We shall return to the problem of quasars in a later chapter. For the present we concentrate on the

Figure 4.19 Martin Rees, a theoretical astrophysicist based in Cambridge, predicted superluminal quasar jets and suggested that black holes are the source of the quasars' unimaginable power. [Courtesy Prof. Sir Martin Rees.]

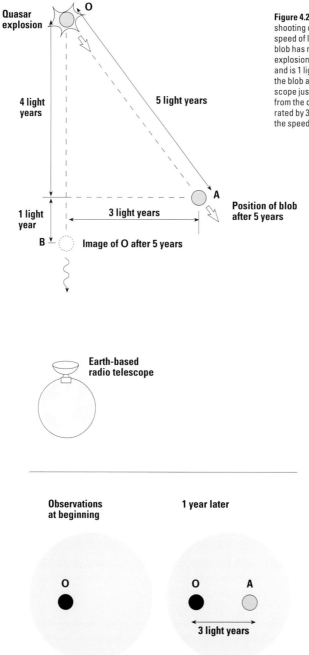

Figure 4.20 The diagram shows a blob of material shooting out of the core of a quasar at almost the speed of light along the path O–A. After 5 years the blob has reached A, while the image of the initial explosion at the core has travelled 5 light years and is 1 light year nearer the Earth. The image of the blob at A will therefore arrive at the radio telescope just 1 year after the image of its creation from the core. Since these two images are separated by 3 light years, the blob *apparently* exceeds the speed of light.

86

question of superluminal sources. In 1966, Martin Rees, a young astronomer from Cambridge University, suggested that explosions might occur in quasars resulting in fragments being ejected at near light speed. Further, he predicted that these fragments would appear to us on the Earth as if they were separating from the quasar core at faster-than-light speeds. His prediction was verified some four years later. To understand the origin of these superluminal speeds, look at Figure 4.20. Suppose a fragment emitted from a quasar has reached position 'A' after five years. Notice that light emitted from the quasar when the fragment emerged will reach Earth only one year before radiation emitted from the fragment at position 'A'. In this example we see that the fragment appears to be receding from the quasar at three times the speed of light. Other superluminal jet speeds can be explained in a similar way.

So, is there any hope for 'genuine' faster-than-light travel? Some physicists have suggested that there may be elementary particles capable of moving faster than light. The US physicist Gerald Feinberg has dubbed such particles 'tachyons' from the Greek word for swift, 'tachus'. Most physicists will say severely that they believe tachyons to be a 'very speculative' idea. Certainly, no one has ever found any evidence for such incredible particles, and it is not clear that one can devise a fully consistent relativistic theory including them. In the next chapter we go on to consider the implications of Einstein's relativity for Newton's laws of motion. We shall see that there are good reasons for believing that the speed of light represents the ultimate speed for matter.

$E = mc^2$

The most important result of a general character to which the special theory of relativity has led is concerned with the conception of mass. Before the advent of relativity, physics recognized two conservation laws of fundamental importance, namely, the law of conservation of energy and the law of conservation of mass; these two fundamental laws appeared to be quite independent of each other. By means of the theory of relativity they have been united into one law.

Albert Einstein, *Relativity*, 1916

Phlogiston and caloric

Before we look at Einstein's famous equation, we had better set the scene by describing how scientists had arrived at the concepts of energy and mass – the 'E' and 'm' in the equation. The gradual evolution of our present understanding of energy began with two ideas at least as curious as the infamous aether – 'phlogiston' and 'caloric'. It provides us with some insight into the way that science progresses to look at why these two theories were invented and subsequently discarded.

Phlogiston was introduced towards the end of the seventeenth century by Georg Stahl, a German professor of medicine and chemistry, in an attempt to understand fire. Even in the latter part of the eighteenth century, many scientists still regarded fire as an element. Combustible materials were supposed to be made up of two parts – the calx, or ash, and the 'phlogiston'. It was thought that, when a substance burned, the phlogiston was liberated, leaving the ash behind. Different substances contained different ashes, but phlogiston was assumed to be common to all elements that could be burned. The air in which these substances burned also appeared to be a very confusing substance. Air could apparently take up a certain amount of phlogiston – but, when completely 'phlogisticated', could no longer sustain combustion. When it was pointed out that some materials, such as tin and lead, actually gained in weight when burnt to ash, a new property of phlogiston was proposed. Phlogiston was supposed to add 'levity' to the metal – this then 'explained' how the ash could be heavier than the metal when the phlogiston

Figure 5.1 Antoine Laurent Lavoisier (1743–1794) pictured with his wife, who became his partner for his chemistry investigations. Lavoisier became wealthy because of his appointment to the French government's 'Fermes Generales' – an organization that paid the government a fixed sum for the privilege of collecting and retaining taxes. From 1776, Lavoisier worked at the French Royal Arsenal, in charge of gun-powder production. Although he was a liberal who sympathized with the French Revolution in its early days, his involvement with the corrupt system of 'tax farming' led to his execution by guillotine. Appeals from fellow scientists for a reprieve were dismissed by the judges with the statement 'The Republic has no need of savants.' His wife survived him, and later married the physicist Benjamin Thomson, Count Rumford. [The Metropolitan Museum of Art, Purchase, Mr and Mrs Charles Wrightsman Gift, in honour of Everett Fahy, 1977 (1977.10).]

was removed by burning in air. Over the years, the phlogiston theory had to become more and more ingenious to explain all the experimental findings. The person who was most responsible for clarifying the discrepancies between theory and experiment was the French scientist Antoine Laurent Lavoisier. He wrote:

> Chemists have turned phlogiston into a vague principle…which conse-quently adapts itself to all the explanations for which it may be required. Sometimes this principle has weight, and sometimes it has not; sometimes it is free fire and sometimes it is fire combined with an earthy element; some-times it passes through the pores of vessels, sometimes these are impervious to it… . It is a veritable Proteus changing in form at each instant.

The slippery nature of the phlogiston theory is reminiscent of the desperate attempts that were made to preserve the luminiferous aether.

In the eighteenth century, there was great confusion about the nature of electricity and magnetism, and also about heat. Scientists were fond of introducing 'fluids' to explain these phenomena. Some scientists thought of electricity in terms of two fluids, vitreous and resinous, with two more fluids, austral and boreal, describing magnetism. Heat, on the other hand, was envisaged as a single fluid called 'caloric'. Hotter bodies were supposed to contain more caloric than cooler ones. Heating a body therefore involved caloric flowing to that body from somewhere else: the total amount of caloric was believed to remain constant. Lavoisier, who did so much to destroy the phlogiston theory of fire, included caloric in his list of chemical elements published in 1789.

Not all scientists believed in these fluids. Almost one hundred years before, Newton had arrived at different, more tentative, answers to these questions concerning the nature of fire and heat. He believed that God had made indestructible atoms when he created the universe, and that each atom had a definite mass and the ability to attract or repel other atoms by means of forces. This suggested that the basic properties of matter were mass and charge. Electric forces act between stationary charges: as we now know, magnetic effects come into play for moving charges. Newton believed that fire was the production of large amounts of light accompanying a chemical reaction. In this suggestion, Newton was ahead of his time and almost alone in his belief. In fact, Newton was well aware how difficult it was to reconcile his ideas with the many confusing results obtained by the chemists.

Caloric was not the only proposal for a theory of heat. Another view of heat had been developed almost a century earlier, nowadays called the 'kinetic theory' of heat. This held that heat was due to the motion of the molecules that made up the substance, so, in the case of a gas, the hotter it was, the faster the motion of its molecules. In 1738, the famous French scientist Daniel Bernoulli was able to account for many of the properties of gases in terms of gas molecules in motion that were able to collide with each other and with the walls of the vessel in which they were contained. Despite the successes of this kinetic theory, the caloric theory of heat was almost universally accepted by the late eighteenth century.

The demise of the phlogiston theory began with a discovery by Joseph Priestley, who, curiously enough, remained a staunch supporter of phlogiston to the end of his life. Priestley found that air was not a simple substance but was made up of several different gases. He further discovered that only one of these components of air supported combustion. Since the term 'phlogisticated air' was used for air that was no longer able to support combustion, Priestley called this new substance 'dephlogisticated air'. Lavoisier called this same component 'respirable air' or 'vital air', since Priestley had shown that animals use up this dephlogisticated air when they breathe. Today we call this oxygen. Priestley also showed that green plants regenerate oxygen when exposed to sunlight. It was after Priestley visited him in Paris in 1774 that Lavoisier carried out the series of experiments that enabled him to unravel the confusing results of the chemists.

Lavoisier heated mercury in a closed vessel and showed that it burned in air to form a red calx or ash, gaining weight in the process. The air left behind in the vessel extinguished lighted candles and did not support animal life. We now understand that oxygen had been removed from the air to form red mercuric oxide. Lavoisier showed that, for the closed system, there was no overall change in weight before and after this process. Instead of phlogiston being removed by heating, oxygen had been 'added' to the metal to form an oxide of mercury. Strong heating to higher temperatures then caused the calx to decompose and the bound oxygen to be set free. When this

Figure 5.2 One of Lavoisier's experiments, in which mercury in a glass balloon was heated over a furnace. All the gases produced were captured in a bell jar placed in a trough containing mercury. [Ann Ronan Picture Library.]

Figure 5.3 A cartoon of (Sir) Benjamin Thomson (1753–1814), also known by his Bavarian title of Count Rumford. Thomson was an American, born in New England, but thought it prudent to flee the country during the American Revolution because of his close association with the British. During his travels in Europe, he once met Edward Gibbon, who was then writing his famous book *The Decline and Fall of the Roman Empire*. Gibbon described Thomson as 'Mr. Secretary-Colonel-Admiral-Philosopher Thomson'. This description fits well with Thomson's activities during the eleven years he spent in Bavaria – he was Minister of War, Minister of Police, Grand Chamberlain and scientist. This cartoon shows him standing in front of a 'Rumford Stove', on top of which are two pots, also of his own design. [By courtesy of the Royal Institution.]

released gas was mixed with the oxygen-free air left over from the first experiment, normal breathable air was reproduced. Lavoisier's experiments were the first to identify clearly the role of oxygen in combustion, and they dealt a fatal blow to the phlogiston theory.

What of the caloric theory of heat? To this day, when we talk of 'sunbathing', we are in fact using the metaphor of a fluid for heat. We use the same metaphor when we talk about heat 'flowing' from a hotter body to a cooler one. In spite of his pioneering work on combustion and his subsequent demolition of the phlogiston theory, Lavoisier, like most scientists of his time, still believed in the caloric theory of heat as a substance. The decline of the caloric theory really began with some interesting experiments following on from those of Lavoisier. These were performed by another curious character in the history of physics. This was the American Benjamin Thomson, who later took the title Count Rumford when he was elevated to the nobility of the Holy Roman Empire. The experiments carried out by Thomson were performed in his capacity as the Bavarian Minister of War. After supervizing the boring of a cannon he said:

> I was struck by the very considerable degree of heat which a brass gun acquires, in a short time; and with the still more intense heat of the metallic chips separated from it by the borer.

Horses working the borer against the resistance of the metal caused heat to be produced at a steady rate: the supply of heat seemed limitless. He concluded that

> any thing which any insulated body, or system of bodies, can continue to furnish without limitation, cannot possibly be a material substance.

Thomson also demonstrated that the weight of a gold sphere was not measurably changed by heating. This showed that, if heat was a substance, then the mysterious caloric must be almost weightless. These and other experiments all suggested that heat was better interpreted as some form of internal motion – as proposed long before by the kinetic theory. It now seems surprising that all these negative results did not immediately lead to the end of the caloric theory. In fact, one reason was the lack of understanding of the physical heat of a hot object and the 'radiant' heat given off by such a body. We now know radiant heat to be a form of electromagnetic radiation, but without knowledge of such a radiant component of heat, the kinetic theory could not explain the transfer of heat through a vacuum. So it was that, in spite of Thomson's challenging results, most scientists still preferred to believe in a theory which appeared to require every substance to contain unlimited amounts of weightless caloric.

Energy and atoms

The caloric theory was eventually abandoned in the middle of the nineteenth century. This came about as a result of a series of painstaking experiments that demonstrated beyond doubt that different forms of energy could be converted into each other but that the total amount of energy remained the same. Although the physicist James Prescott Joule was responsible for performing the key experiments that led to the principle that we now know as 'conservation of energy', another lesser-known scientist deserves a mention. Julius Robert Mayer was a German doctor serving on board a ship in the tropics. He was puzzled by the fact that, in contrast to his previous observations back in temperate Germany, the blood from the veins of patients in the tropics was almost as red as the blood from the arteries. Mayer believed that blood was responsible for taking oxygen around the body, providing the essential element needed for the slow transformation of food into body heat. To explain this difference in colour, Mayer suggested that the high temperatures of the tropics meant that less oxygen was required by the body to produce heat – leaving a large amount of unused oxygen in the returning veinous blood. In addition, Mayer was aware that strenuous exercise or work makes us hot. Mayer thought deeply about these observations, and concluded that the amount of heat generated was directly related to the amount of work expended. He wrote: 'force, once in existence, cannot be annihilated; it can only change its form.' We would now use the word 'energy' in place of Mayer's 'force' – 'Kraft' in German, but it is clear that Mayer had, in fact, arrived at the idea of energy 'conservation'. His conclusions were published in Germany in 1842, and were largely ignored by the scientific establishment. Possibly as a result of this neglect, Mayer suffered a breakdown in his health – a not uncommon story in the annals of science.

Figure 5.4 Julius Robert von Mayer (1814–1878) was the first to formulate the law of conservation of energy. Mayer's work was rescued from obscurity by the efforts of the British physicist John Tyndall. As a result, he gained some recognition before he died and was granted the right to use 'von' before his name, the German equivalent of a British knighthood. [The Mansell Collection.]

The scientist who is now credited with establishing the principle of energy conservation is James Prescott Joule. Joule was the son of a wealthy Manchester brewer, and he sent his first scientific paper to the Royal Society at the age of twenty-three. In this paper, Joule showed that an electric current flowing through a conductor produced heat at a rate proportional to the

Figure 5.5 A painting of the pithead of a mine in about 1790. The concept of energy grew out of meditations on the operation of machines created in the industrial revolution. [The Board of Trustees of the National Museums & Galleries on Merseyside (Walker Art Gallery, Liverpool).]

resistance of the conductor times the square of the current. The heat generated by this effect is still called Joule heat. Joule then went on to look at the changes in energy that were required to produce this heat. In his experiments, the electric current originated from a chemical reaction in batteries. Chemical energy is converted into electrical energy in the batteries, which finally produces heat in the wire of the conductor. Joule is best known for measuring what is called the 'mechanical equivalent' of heat – how much work generates how much heat. He showed that a given amount of mechanical work – used to run a dynamo to produce an electric current – is converted into a definite amount of heat in the conducting wire. His most famous experiment showed that the same amount of heat can be produced directly from the same amount of mechanical work without going through the intermediate conversion into electrical energy. In this experiment, Joule measured the amount of heat produced in water by rotating paddles driven by falling weights. At first, Joule's discoveries went almost as unnoticed as those of Mayer but, at the 1847 meeting of the British Association, Joule was permitted to make a short presentation. Fortunately for Joule, the young and gifted scientist William Thomson was in the audience. Thomson immediately recognized the importance of Joule's work and championed his cause. William Thomson later became Lord Kelvin, and is acknowledged as one of the founders of the modern theory of thermodynamics. After Thomson's endorsement, Joule received his fair share of fame and glory – and a

Figure 5.6 Hermann von Helmholtz (1821–1894) was twenty-six years old and working in medicine when he published his ideas on the conservation of energy. He went on to do important work in both physiology and physics, and he became the leading German scientist of his day. [Deutsches Museum, München.]

memorial in Westminster Abbey after his death.

A full appreciation of the work of both Mayer and Joule was finally achieved with the publication of a paper by the German scientist Hermann von Helmholtz. The paper was entitled 'On the conservation of force (Kraft)'. The modern term 'energy' was introduced later by the Scottish engineer William Rankine. The idea that energy was constant and not destroyed enabled a whole range of previously separate parts of science – chemistry, electricity, magnetism, optics and heat – to be inter-related. Each area displayed a different form of energy. These different forms could be transformed into each other by suitable experimental arrangements, but the total amount of energy always remained the same. A simple example of energy transformation and conservation occurs on a roller-coaster: the gravitational potential energy of the carriage at the top of a hill is rapidly turned into kinetic energy as it descends.

In the late nineteenth century, all of science appeared to be on the brink of being fully understood. Not only was energy conserved and indestructible, but so too was matter itself. Most scientists accepted that the creation and destruction of atoms was impossible. The importance of atoms in chemistry had first been demonstrated by a Manchester schoolmaster named John Dalton in 1808. In order to understand Dalton's idea, let us consider an example. He found that the weight of oxygen needed to transform a certain amount of carbon into 'carbonic acid' was twice as much as was required to make 'carbonic oxide'. The modern names for these two compounds incorporate Dalton's atomic model for them – carbon dioxide (CO_2) and carbon monoxide (CO), respectively. A carbon dioxide molecule contains two atoms of oxygen attached to each carbon atom, whereas a carbon monoxide molecule has just one atom of each element. Following on from these early experiments of Dalton, the idea that atoms could not be created or destroyed in chemical reactions became an established principle of chemistry.

Newton meets Einstein

Newton's laws of motion had served science well for over 200 years. His three laws had enabled physicists to gain a quantitative understanding of a vast range of phenomena – from the motion of planets to the motion of molecules in the kinetic theory. But, as we saw in the last chapter, Einstein's removal of the notion of an absolute time and the curious behaviour of the speed of light lead us inexorably to the conclusion that nothing can be accelerated to speeds faster than light. This is in clear contradiction of Newton's laws of motion. Before describing a 'thought experiment' that demonstrates Einstein's mass–energy relation, we first describe an experiment with electrons that shows Newton's laws to be inadequate at high speed.

Electrons may be accelerated by an electromagnetic field. Such a field may be generated by charging two metal plates with opposite electric charges so that there is an electrical 'potential difference' between the plates. This potential difference is defined in terms of the energy required to move a charged particle from one plate to the other against the force of the electric field between the plates. The unit used by scientists to quantify such electric 'potential differences' is called the 'volt', after the Italian physicist Alessandro Volta. In 1800, Volta had developed the first practical electric battery – the 'voltaic pile' – which opened the way for the detailed investigation of electricity and magnetism by such physicists as Davy, Faraday, Ohm and Ampère. In dealing with the physics of atoms, a convenient energy scale may be defined in terms of the amount of energy gained by an electron when it 'falls' through a potential difference of 1 volt. Compared with the energy required to heat up a kettle of water, this is a tiny amount of energy, but one much more appropriate to atomic energy scales than the quantities of energy defined by the 'calorie' and 'joule' units appropriate to everyday life. For this reason, this 'atomic-sized' amount of energy is called an 'electron-volt', or 'eV' for short. Because of their tiny mass (electrons carry the same, but opposite, electric charge as a proton but are some 2000 times lighter), electrons can easily be accelerated to high speeds.

The results of an experiment to measure how the speed of an electron varies with the amount of energy given to the electron are shown in Figure 5.7. According to Newton, the energy of the electron is proportional to the square of its velocity. As can be seen from the graph, the experiment clearly shows that the Newtonian prediction – marked 'Newton' on the figure – is wrong. As the energy of the electron increases, instead of the speed of the electron exceeding the velocity of light, the measured speed gets closer and closer to the speed of light but never quite reaches it. The experimental points lie on the curve labelled 'Einstein', which is the prediction of special relativity for this experiment. This experiment, which is typical of those carried out daily at high-energy particle accelerators throughout the world, shows that Newton's laws of motion break down at high speeds. Is there any escape from this conclusion? One possible escape route could be our assumption about the energy of the electron. Perhaps the electron is not being given all the energy corresponding to the accelerating voltage? We can test this by measuring

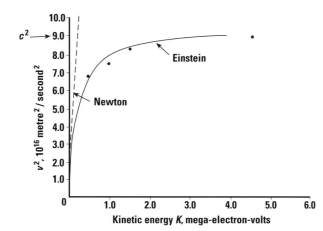

Figure 5.7 The graph showing the results of an experiment that measures the speed, *v*, of the electrons as the kinetic energy of the electron, *K*, is varied. The Newtonian prediction is the dashed line marked 'Newton', whereas the data clearly follow the prediction of relativity shown by the solid line marked 'Einstein'. The dashed line at 9.0×10^{16} metre²/second² corresponds to the square of the velocity of light, c^2.

the energy of the electron directly. In the particular experiment shown here, this was done, and the result clearly confirmed that the electrons have the full energy. There is no escape: Newton's laws must be modified to describe particle motion at high velocities.

What could be wrong with Newton's laws of motion? To answer this question we return to Einstein's favourite device – the thought experiment. For simplicity, we shall dispense with any attempt at realism, although there is nothing in principle to prevent us carrying out such an experiment. Rather than talk about rocket ships moving at near light speeds, we return to the more familiar example of a train. The participants in our experiment are two skillful basketball players. Figure 5.8 will illustrate this example. One player is standing at rest by the railway track; the other is on the train opposite an open window. Both players are holding identical basketballs. They then carry out the following experiments. In the first experiment the train is stationary so that the two players are both standing at rest opposite one another. At the same moment they each throw their basketballs towards the other with exactly the same speed and trajectory. They aim the basketballs in such a way that the balls collide halfway and each ball returns to the hands of the player that threw it. The balls return on the same path as they followed on their outward trip. There is nothing surprising in this: it is exactly what we expect from the law of 'conservation of momentum'.

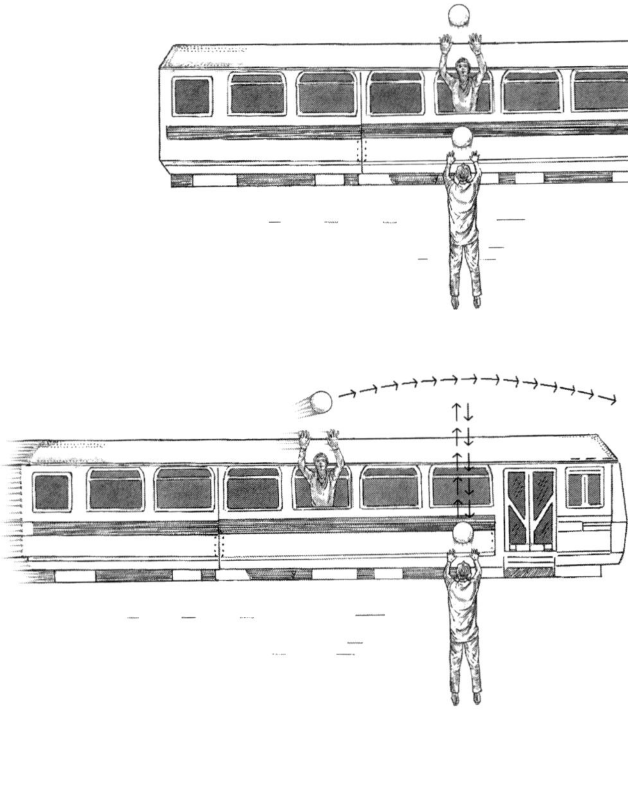

Figure 5.8 The top picture shows a basketball player on a stationary train and a colleague on the platform, each throwing a basket ball, which subsequently collide and bounce back to be caught by the players. The bottom illustration shows the more difficult experiment performed from the window of a rapidly moving train. The lines of arrows show the trajectory of the balls. Notice that the ball thrown from the moving train continues to move rapidly in tandem with the train. However, from the point of view of the stationary player his colleague seems to throw the ball more slowly because of 'time dilation'. (Relativistic shortening of the train has been ignored.)

Conservation of momentum follows directly from Newton's laws of motion. Its consequences are familiar to us from collisions of everyday objects. Newton's 'quantity of motion' – now called momentum – is the mass of a body times its velocity. A heavy truck moving at 50 miles/hour has more momentum than a small car moving at the same speed – and is consequently harder to stop in a collision. The truth of Newton's conservation of momentum is evident from many examples in everyday life. It applies in collisions of snooker balls: the combined momentum in each direction remains the same before and after impact. It also applies to situations such as firing a gun. Before the trigger is pulled, both gun and bullet are stationary: after firing the gun, the bullet, with small mass but high velocity, rushes forward out of the gun barrel. This momentum change of the bullet is balanced by the recoil of the gun, which has a much larger mass and a much smaller velocity. Although we have presented the idea of conservation of momentum as 'obvious' – familiar from our everyday experience – it is amusing to consider that, in the 1920s and 30s, the idea of rocket ships operating in space was controversial. One of the pioneers of space flight was ridiculed by the *New York Times* in a famous editorial:

> That Professor Goddard, with his 'chair' in Clark College and the countenancing of the Smithsonian Institution, does not know the relation of action to reaction, and of the need to have something better than a vacuum against which to react – to say that would be absurd. Of course he only seems to lack the knowledge ladled out daily in high schools.

Not surprisingly, Goddard became withdrawn and reticent about publicizing his ideas. It was a great irony that the US Government, after refusing to acknowledge its infringement of hundreds of Goddard's rocket patents, was obliged to pay his widow one million dollars in compensation, fifteen years after Goddard's death. The NASA–Goddard Space Flight Center at Greenbelt, Maryland just outside Washington, D.C., is named in his honour.

Let us return to our two basketball players. We now understand what happens when both are at rest. The surprise in our thought experiment comes when the two players try to repeat their trick when the train is moving at high speed. The player on the train just has to repeat his throw exactly as he did when the train was stationary – but now both players must time their throws just right to allow for the sideways motion of the train. According to Newton, this experiment could be done, with conservation of momentum guaranteeing that each player gets his basketball back. So what goes wrong if the train could move with 'relativistic' speeds? Consider the collision as viewed by each player. The player by the track sees the player on the train moving past at high speed. According to him, he sees the player on the train throw the ball out more slowly than he did when they were both at rest. This follows either from the formula for the relativistic addition of velocities that we discussed in the last chapter or directly from a time dilation argument. The player on the train, on the other hand, sees his friend standing by the track, approaching at high speed. He throws his ball out of the window with just the same speed as before, but he now sees his friend throw his ball towards the side of the train more slowly than before. The result is the same no matter whose point of view we take. In the first experiment, when both are at rest, each sees a momentum change of twice the mass of the ball times its velocity, so conservation of momentum holds. In the case of the moving train, momentum conservation seems to be violated: one ball has the same momentum change as before but the other appears to have a smaller momentum change due to the time dilation factor. In order to maintain conservation of relativistic momentum, we are forced to conclude that mass must change with velocity. From our example, we see that momentum conservation can be restored if the mass of the fast moving ball increases by just the same time dilation factor by which the velocity is reduced.

This thought experiment has enabled us to see the requirement for an increase in mass with velocity. Einstein used a different 'gedanken' experiment to derive his famous result: this particular thought experiment was first presented in a paper in 1909 and will be discussed later in this chapter. We can now understand the results of the electron velocity experiment described above. Besides increasing the velocity of the electron, the energy also goes into increasing its mass. Can we measure the predicted mass increase more directly? In 1901, Walter Kaufmann performed some detailed experiments relating the energy of electrons to their velocity, and he showed that there was an apparent variation of mass with speed. Several physicists – notably Max Abraham and Alfred Bucherer – then devised models for this variation in mass based on a model of the electron as a small charged sphere. After the momentous development of relativity in 1905, Kaufmann was stimulated to repeat his experiments with greater accuracy. In 1906 he announced:

> The measurements are incompatible with the Lorentz–Einstein postulate. The Abraham equation and the Bucherer equation represent the observations equally well.

This result caused much excitement amongst the physics community, but Einstein was impressively unmoved. He wrote:

> It is…noted that the theories of Abraham and Bucherer yield curves which fit the data considerably better than the curve obtained from relativity theory. However, in my opinion, these theories should be ascribed a rather small probability because their basic postulates concerning the mass of the moving electron are not made plausible by theoretical systems which encompass wider complexes of phenomena.

Soon after Einstein wrote these words, he received a letter from Bucherer. In the letter, Bucherer told Einstein that new experiments he had performed had led him to abandon his own model in favour of relativity. An updated version of Bucherer's results is shown in Figure 5.9. As can be seen, the agreement with the relativistic mass increase predicted by relativity is excellent.

The need for relativistic momentum conservation with a velocity-dependent mass can be understood in another way. Imagine a typical shot in snooker: the cue ball colliding with another ball at rest. If the balls meet in a slightly off-centre collision, both will move off in different directions. Using Newtonian energy and momentum conservation, it is not difficult to show that, for two balls of equal mass, the angle between the two outgoing balls must be 90 degrees. What happens if the incoming ball is moving with 'relativistic' speed? In this case, although the masses of the two balls were equal when both were at rest, the mass increase of the moving ball gives the cue ball much more momentum than Newton would have expected. This makes the collision look more like the collision of a very massive ball with a much lighter one. The relativistic ball will tend to follow through after the collision in much the same way as a train hitting a car on a railway crossing. The fast ball will hardly be deflected from its path, while the stationary ball will be carried forward. This intuition is borne out by the mathematics, and the angle between the two balls as they emerge from the collision can be much less than 90 degrees. We have described this situation as if it were yet another thought experiment – a relativistic snooker game. In fact, in collisions between elementary particles, such as protons and electrons, we can see relativity in action.

Figure 5.9 A graph showing how the electron mass varies with speed. The graph plots data from several experiments that conclusively show that the mass of the electron increases as the speed of the electron is increased. The *y*-axis is the ratio of mass at velocity *v* to the mass at rest; the *x*-axis is *v* divided by the velocity of light, *c*. The smooth curve through the points is the variation predicted by Einstein.

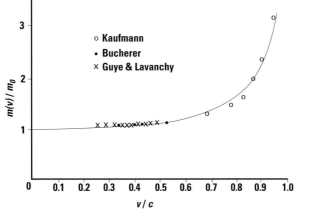

The equivalence of mass and energy

…that light transfers mass. … This thought is both amusing and attractive; but whether or not the good Lord laughs at me concerning this notion and has led me around by the nose – that I cannot know.
Albert Einstein, letter to Conrad Habicht, 1905

In our discussion of the velocity addition law in the previous chapter, we concluded that it was impossible for anything to move faster than the speed of light. Einstein's mass formula gives us another way of looking at this result. As the speed of a particle approaches the velocity of light, its mass increases. We have seen that adding energy to a particle that is already moving at a relativistic speed mainly serves to increase the mass of the particle, without significantly increasing its speed. It was this line of thought that led Einstein to conclude that the mass of a body is a measure of its energy content. There are many ways to arrive at his celebrated equation. We will describe Einstein's original thought experiment.

In order to appreciate Einstein's argument, we must first appreciate that light possesses both energy and momentum. It is natural to think that light carries energy – after all, it is clear that the Earth is warmed by light from the Sun. In 1905, Einstein had used the hypothesis that light could be regarded as a stream of individual packets of energy to explain the 'photoelectric effect' – the fact that light shining on a metal surface can cause electrons to be ejected. Einstein's explanation centred on the idea that the light energy interacted with the electrons bound to the atoms in the metal as if concentrated in particle-like packets of energy. This phenomenon enabled even a weak light source to knock electrons out of the metal surface with no noticeable time delay. In contrast, the wave picture of light predicted that the

Figure 5.10 The original apparatus used by Hertz to discover the photoelectric effect. An electric current was passed across the gap between the electrodes. When the gap was increased the electric current ceased, but when light was shone on the negative electrode the current flowed again. Only light with a frequency greater than a certain minimum frequency will release electrons. This minimum frequency varies with the type of metal. [Deutsches Museum, München.]

light energy would be distributed all over the metal surface, and therefore that electrons would only be ejected when sufficient energy had accumulated after a measurable time delay. The experimental results supported Einstein's particle-like picture for the interaction of light with the electrons in the metal. The name 'photon' that we now use to refer to these particles of light was introduced around twenty years later by the Berkeley chemist Gilbert Lewis.

We have seen that it appears to be impossible to accelerate a massive particle up to the speed of light, since as we increase the energy in an attempt to accelerate the particle to a higher speed, we find that the mass becomes larger and larger. Einstein's formula predicts that, at the speed of light, the mass of the particle would become infinitely great. In other words, it would require an infinitely large amount of energy to accelerate the particle to the speed of light. This is, quite obviously, impossible. How then does a photon manage to travel with the speed of light? The answer is that photons have zero 'rest' mass, which means that the energy of a photon is all kinetic energy. If a photon is forced to come to rest in some absorbent material, it simply ceases to exist.

If individual photons have a definite energy and are able to behave like particles, we should expect them to carry momentum. This was anticipated by Maxwell's theory of light. An ordinary beam of light contains huge numbers of photons – 1 watt of visible light corresponds to a flow of around 1 million million million photons/second. The relation between energy and momentum was first predicted by the British physicist John Poynting on the basis of Maxwell's theory. This prediction has been tested by measuring the 'radiation-pressure' of a beam of light using a *radiometer* – a thin metal vane suspended in a vacuum. The first experiment conducted to measure radiation-pressure was performed by the Russian physicist Peter Lebedev in 1901, but care had to be taken to separate the effect due to true radiation-pressure from the potentially confusing 'radiometer effect'. This is the phenomenon that drives the toy radiometers sometimes sold as novelties. These radiometers rotate in the *opposite* direction to that expected when considering radiation-pressure alone. The vanes of the radiometer are painted black on one side and white on the other. The black side absorbs more photons and heats up more than the white side, which is more reflective. Since the momentum change caused by a reflected photon is twice as much as that caused by an absorbed one, the radiation-pressure will tend to turn the device in the direction from white to black. Unfortunately, this effect is swamped by local heating of the residual gas in the radiometer. The gas is hotter close to the black side, and its faster-moving molecules turn the radiometer in the opposite direction. A more modern demonstration of the radiation-pressure exerted by light has been performed using a laser beam to levitate a small glass sphere.

Lebedev also suggested that radiation-pressure was the reason why the tails of comets always point away from the Sun. In fact, a stream of fast-moving particles originating from the Sun, called the solar wind, causes considerably more pressure than sunlight. This solar wind causes the dust, molecules and atoms in the tail of a comet to be blown in a direction away

Figure 5.11

A photograph of the comet Hyakutake taken in April 1996 from the Izana Observatory in Tenerife. Undergraduates from the Physics and Astronomy Department at the University of Southampton carrying out projects at the observatory were fortunate to have a grandstand view. [Photograph by Malcolm Coe.]

In March and April of 1997, Comet Hale-Bopp became one of the most spectacular and visible comets this century. It was discovered on the night of July 22nd, 1995, by two astronomers: Dr Hale, who has a degree in astronomy, with his telescope in Cloudcroft, New Mexico, and Mr Bopp, an amateur astronomer with the North Phoenix Alternative Astronomical Society, 400 miles away in Arizona. After their independent discovery of the comet had been confirmed, Alan Hale telephoned Tom Bopp and introduced himself with the words 'I think that we have something in common.' [Photograph by Tim Printy. Taken on 5 April 1997. Exposure time was 8 minutes using a 200 mm F2.8 lens on Kodak PJM film.]

Figure 5.12 An artist's impression of a solar sail destined for Mars or the Moon. [Cambridge Consultants Ltd.]

from the Sun. Another Russian, the space pioneer Konstantin Tsiolkovski, together with fellow countryman F. A. Tsander, put forward the idea that a spacecraft could be powered by sailing on the light from the Sun. In a rather more practical demonstration, the effects of radiation-pressure on the US balloon satellite Echo 1, launched in 1960, caused the height of the balloon to vary by as much as 500 kilometres.

Armed with the knowledge that photons possess both energy and momentum, we are, at last, able to appreciate Einstein's original thought experiment. His experiment is made more realistic if we translate it to the *Starship Enterprise*. Imagine the *Enterprise* is floating freely in space and that we are in the shooting range on board ship. If we fire a bullet at a target at the far end of the range, the recoil of the gun required by momentum conservation is ultimately transmitted, via our hand and body, to the whole spacecraft. This causes the ship to move very slowly in the opposite direction to the bullet. When the bullet hits the target at the far end of the range, its momentum cancels out the ship's momentum and the ship comes to a stop. Although the ship has moved, because the mass distribution has been slightly changed (the bullet is at the other end of the range), the centre of mass of the spaceship has remained in the same place. This is what we would expect since the system (the spaceship and the shooting range) is completely isolated in space. Einstein's idea was to use a photon as the bullet. If we arrange that the photon from our laser pistol has the same momentum as the original bullet, momentum conservation will apply in exactly the same way as before and the entire ship will feel the recoil. Because the photon travels faster than the bullet, the spaceship will not have time to travel so far – but it will still have moved. Since the ship is isolated, we again expect that the centre of mass must be unchanged. But the photon has no rest mass, only momentum and energy. In order to keep the centre of mass of the spaceship in the same position, mass must be transferred from one end of the range to the other. We must conclude that the energy transferred by the photon must have an equivalent mass associated with it. With this very simple thought experiment Einstein was able to deduce his famous relation between mass and energy.

Although Einstein's mass–energy relation has dramatic consequences in nuclear physics, as we shall see in the next chapter, it is important to realize the generality of the result. In other words, any change in energy corresponds to a change in mass. A golf ball in motion has more mass than a golf ball at rest. A hot gold sphere has more mass than a cold one – despite Rumford's early measurements. Why did such effects go unnoticed for so long? The reason is that the mass associated with a given amount of energy is extremely small – since the velocity of light is very large. For a 1000 kilograms of gold heated through 1000 degrees, the mass increase is only about one-millionth of a gramme – well beyond the accuracy of Rumford's experiment, and well outside the range of our everyday experiences.

6   Matter and anti-matter

*It is possible that radioactive processes may become
known in which a considerably larger percentage of the
mass of the initial atom is converted into radiations of
various kinds than is the case for radium*

Albert Einstein, 1907

Prologue

In this chapter, we shall explore the application of special relativity to atomic
physics. In order to make accurate quantitative predictions for the atomic
structure underlying both physics and chemistry, it turns out to be essential
to take into account 'relativistic corrections' to the standard quantum
mechanical picture of the atom. Such applications of relativity form an
important part of the experimental evidence supporting Einstein's theory,
and to ignore them would give a distorted impression of its success. Thus,
although this is a book about relativity, in order to appreciate this area
of applied relativity, it is necessary to give a brief overview of our present
understanding of atomic physics. A companion volume, *The Quantum
Universe*, gives a fuller account of the development of quantum theory and
its application to the modern world.

    We begin our overview with an account of the great debate about
the existence of atoms. In 1905, the same year that he published his paper
on special relativity, the young Einstein made a crucial contribution to the
debate with an atomic explanation of Brownian motion. At the same time as
this debate was raging, the first discoveries about radioactivity were being
made by Roentgen, Bequerel and the Curies. A major puzzle of the era was
the origin of the energy released in radioactive decays. Radioactivity was
exploited in the pioneering work of Rutherford, Bohr and Chadwick in their
exploration of the nuclear atom. At about the same time, it also became clear
that there are strong and weak nuclear forces, in addition to the familiar

forces of gravity and electromagnetism. During the 1920s, Bohr's primitive quantum 'recipes' were replaced by modern quantum mechanics, which put our understanding of the atom on a new and radically different footing. Although the main features of atomic spectra could now be explained, it was not until relativistic corrections were incorporated into the theory that the observed 'fine structure' could be understood. This crucially involved a new property of the electron known as 'electron spin'.

Electron spin remained something of a mystery until 1928 when a young Englishman named Paul Dirac produced an equation describing the relativistic quantum mechanics of the electron. The Dirac equation not only correctly reproduced the fine-structure results, previously grafted on to quantum mechanics in a rather *ad hoc* way, but also predicted a wholly unexpected new type of matter. In addition to the electron, Dirac's equation predicted that there should exist an 'anti-particle' partner to the electron. This is now known as the positron, and it was discovered experimentally in 1932. When an electron and a positron meet, they annihilate each other and the energy resulting from their disappearance re-appears as energetic photons. The successful prediction of anti-matter was a spectacular triumph for special relativity. Besides particle–anti-particle annihilation, particle–anti-particle pairs can be created from energy; this is a very visible manifestation of Einstein's mass–energy equivalence.

The purpose of this chapter is to spell out the above story in some detail. In chapter 7, the focus turns to the nucleus rather than the atom when we look at the application of special relativity to nuclear physics. Those readers eager to follow the story of relativity can safely by-pass the remainder of this chapter and all of the next without missing any key steps in the logic. What may be lost is a full appreciation of the part played by special relativity in our present understanding of atomic and nuclear physics.

Atoms are reversible

By and by I despaired of the possibility of discovering the true laws by means of constructive efforts based on known facts. The longer and more despairingly I tried, the more I came to the conviction that only the discovery of a universal formal principle could lead to assured results. The example I saw before me was thermodynamics. The general principle was there given in the theorem: the laws of Nature are such that it is impossible to construct a perpetuum mobile (of the first and second kind).
Albert Einstein, *Autobiographical notes*, 1949

Today, the atomic basis of matter is taught routinely in schools so it is difficult to imagine the suspicion and hostility at the end of the last century to the idea that atoms existed. This hostility seems curious, given the fact that the idea of atoms as eternal, elemental constituents of matter dates back to the Greek philosophers Leucippus and Democritus in the fifth century BC. Centuries later, Isaac Newton wrote: 'It seems probable to me that God in the beginning formed matter in solid, massy, hard, impenetrable, movable particles...'. Over 150 years later, James Clerk Maxwell modified Newton's 'action at a distance' force laws by introducing the idea of the electromagnetic field, but he still remained convinced of the existence of atoms. So why was the atomic theory not a generally accepted part of the scientific establishment by the end of the nineteenth century?

Part of the problem was the inaccessibility of atoms and consequently the lack of direct experimental evidence. There was no doubt that the *concept* of atoms had proved itself useful in chemistry – as John Dalton had demonstrated in 1808. The same concept had also proved useful in physics: as early as 1738, Daniel Bernoulli had been able to use a 'kinetic theory' of atoms to explain many of the properties of gases. The theory was called 'kinetic' because it accounted for these properties – such as pressure and viscosity – in terms of the motion and collisions of atoms or molecules. Such a mechanistic view of gases is now accepted almost without question. Why then, in 1894, could Robert Cecil, Marquis of Salisbury and former British Prime Minister, address the British Association with the words:

> What the atom of each element is, whether it is a movement or a thing, or a vortex, or a point having inertia, whether there is any limit to its divisibility..., whether the long list of elements is final, or whether any of them have a common origin, all these questions remain surrounded by a darkness as profound as ever.

One of the major arguments used against the atomic hypothesis had its roots in 'Thermodynamics'. Thermodynamics, together with the 'Mechanics' of Newton and the 'Electromagnetism' of Maxwell and Faraday, formed the basis of nineteenth century 'classical' physics. Thermodynamics is concerned with processes that cause energy changes as a result of heat flowing into or out of a system, together with work that is done by or on the system. A simple example of a thermodynamic system is that of gas contained in a cylinder with one end closed and the other end sealed with a movable piston. The gas can do work by expanding and moving the piston. Alternatively, this system can have work done on it by pushing down on the piston and compressing the gas. Heat can also be transferred to and from the gas to change its temperature. Petrol engines, steam turbines and jet engines are all 'heat engines' containing a thermodynamic system that converts heat into mechanical work. Similarly, a refrigerator or 'heat pump' is a thermodynamic system designed to transfer heat from a cold body to a hotter one. Thermodynamics is remarkable in that its predictions can all be derived from two basic 'laws' without regard to any detailed model of matter. Thermodynamics applied to gases, for example, makes reference only to 'macroscopic' quantities like pressure, volume and temperature. On the other hand, the kinetic theory of gases can be used to make the same predictions as thermodynamics, but only at the expense of introducing a detailed 'microscopic' model of a gas as a collection of colliding atoms and molecules.

The two basic laws of thermodynamics seem to be a summary of commonsense experience. A popular way of expressing the laws in the nineteenth century was in terms of the impossibility of creating 'perpetual motion' machines. The first law is just a statement about conservation of energy; it says that it is impossible to build an engine that creates indefinite amounts of energy. This was called 'perpetual motion of the first kind'. The second law is more subtle to understand, but is equally rooted in everyday experience. Suppose we start with a cylinder of compressed air at some temperature. If we let the air expand, this system can be used to drive a piston and do work. After the expansion we find that the air is now cooler than before. We can warm up the cylinder and heat the air back to its original temperature by putting the cylinder in contact with a large source of heat – such as a tank, or 'reservoir', of hot water. At this point, it looks as if the overall result has been to do work with the compressed air and then take heat out of the hot-water reservoir. But we have not yet returned the system back to where it started. This is necessary if we want to build a perpetual motion machine. To complete the cycle we must now do work on the system to

Figure 6.1 This proposed self-blowing windmill is an example of a perpetual motion machine. The laws of thermodynamics forbid such machines, but this has not stopped innumerable inventors from trying to evade these laws.

Figure 6.2 A Newcomen engine consisted of a cylinder enclosing a piston attached to one side of a cross-beam, with the other side attached to the plunger of the water pump. The weight of the pump lifted the piston, and steam entered the cylinder. When water cooled the cylinder, a partial vacuum was created which lifted the pump plunger. Newcomen (1663–1729) lived and worked in Dartmouth in Devon. His engine was the precursor of the modern steam engine, which provided the power for the Industrial Revolution in Britain. [Dartmouth Museum.]

Figure 6.3 Rudolf Clausius (1822–1888) was the son of a Prussian pastor who also ran a small school, which his son attended. In the 1870 war, Clausius was badly wounded while running an ambulance service which he and his students had set up. Clausius is best remembered for his formulation of Carnot's ideas about heat engines, in which he introduced the concept of 'entropy'. [The Mansell Collection.]

Figure 6.4 The tomb of Austrian physicist Ludwig Boltzmann (1844–1906). The equation he discovered relating entropy S to a measure of the disorder W ($S = k \log W$) is engraved on the tombstone. The constant k is named Boltzmann's constant in his honour. The honour was posthumous, since Boltzmann committed suicide after being depressed by years of battles with Mach and his followers about his use of the atomic hypothesis. [Österreichische Nationalbibliothek/Bildarchiv – Wien.]

recompress the air back to its starting pressure. This cycle of events embodies the second law of thermodynamics: it is impossible to build an engine that does nothing but take heat from a source at a constant temperature and converts it all to mechanical work. Such an engine was called a 'perpetual motion machine of the second kind'. By the end of the nineteenth century, these two remarkably simple laws had been shown to have an enormous range of applicability. As can be seen from the quotation at the beginning of this section, Einstein was greatly influenced by this formulation of thermodynamics in his search for a framework for special relativity. The two postulates he proposed as the basis for special relativity may be phrased in a very similar manner, namely, the impossibility of detecting uniform motion and the impossibility of exceeding the speed of light.

What was the origin of the conflict between the 'thermodynamicists' and the 'atomicists'? It all had to do with the notion of the 'arrow of time' and the evident irreversibility of everyday events. The German physicist Rudolf Clausius had shown that the second law of thermodynamics could be rephrased in terms of a quantity he called entropy. Although entropy is a thermodynamic quantity that can be defined without the use of an atomic picture of matter, our modern understanding of entropy is an atomic one due to Ludwig Boltzmann. Boltzmann was an Austrian physicist who, with Clausius, Maxwell and the American Josiah Willard Gibbs, pioneered the field of 'statistical mechanics'. Using the atomic model for gases, Maxwell had been able to relate the velocity distribution of the atoms to the temperature of the gas. At the time, no one had been able to observe the effects of atoms directly, so Maxwell's prediction remained unverified by experiment for some sixty years. Boltzmann used statistical mechanics to relate the entropy of the gas to the amount of 'disorder' in the motion of its atoms. Roughly speaking, if the atoms of a gas are all moving more or less in the same direction, then the energy of the gas atoms can be used to do useful work, such as pushing a piston. Heat energy, on the other hand, is a random motion of atoms that cannot be coordinated for organized work. Since heat is always a by-product of real systems performing work, the entropy or disorder of the whole system always increases. The entropy of the universe is therefore always increasing, and there is a clear distinction between 'before' and 'after' or between 'cause' and 'effect'. In this way, the second law appears to give rise to a fundamental asymmetry in the laws of physics – that there is a clearly defined 'arrow of time'.

This increase in entropy with time was at the heart of the conflict. If the atomic view of matter were true, then we should be able, in principle, to understand all the complex phenomena described by thermodynamics in terms of millions and millions of elementary atomic collisions. If we imagine a collision between two atoms as being like the collision between two billiard balls, we see that there is a fundamental reversibility at this atomic level. If we take a video of the two balls colliding, we could run the tape backwards and see another perfectly possible collision. This ability to run the video backwards or forwards is a demonstration that the mechanical laws of Newton do not distinguish a direction in time. In physicists' jargon, we say that Newton's laws are 'invariant under time reversal'. The problem is that everyday happenings generally do distinguish past from future. Friedrich Wilhelm Ostwald, a well-known German chemist, summarized the problem at a meeting in 1895, at which Boltzmann was also present. He said

> … in a purely mechanical world there could not be a before and an after as we have in our world: the tree could become a shoot and a seed again, the butterfly turn back into a caterpillar, and the old man into a child.

In other words, if atoms were a reality, and the physical properties of matter just the result of reversible atomic collisions, why is the world we see around us so evidently *not* reversible? In similar vein, the great German physicist Ernst Mach replied to an attack by Max Planck with the words:

> If faith in the reality of atoms is essential for you, I declare myself liberated from the physical mode of thought, I don't wish to be a true physicist, I renounce all scientific reputation, finally I render immeasurable thanks through you to the community of believers. Freedom of thought seems to me preferable.

Curiously, it was another of Einstein's 1905 papers that gave the first convincing demonstration of the reality of atoms. In 1827, the botanist Robert Brown reported seeing an apparently random motion of small particles - such as pollen grains – when they were floating in water. This 'Brownian' motion was a problem for the thermodynamicists since it seemed possible in principle to construct a perpetual motion machine from these

Figure 6.5 Drawings of Brownian motion based on work by Jean Perrin (1870–1942). The drawings show the successive positions of particles about one-millionth of a metre in diameter moving in a liquid. Observations were made with a microscope every 30 seconds, and the positions were joined by a straight line. The resulting motion is due to the asymmetrical impacts of the liquid molecules propelling the particle in a random manner.

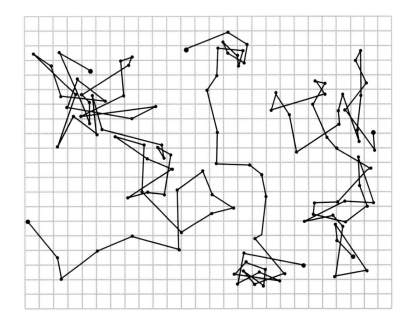

tiny movements. Einstein explained the motion of the pollen grains in terms of collisions with the water molecules, and he made detailed predictions which were later confirmed by experiment. These results, together with the discovery of the electron in 1897, were enough to convince even hardened atomic sceptics such as Ostwald, who, in 1908, conceded that the results 'entitle even the cautious scientist to speak of an experimental proof for the atomistic constitution of space-filled matter'. Mach, however, died unconvinced in 1916. Ten years before, Boltzmann, depressed by years of scientific battles, had committed suicide.

We now understand the second law of thermodynamics in a probabilistic sense. Maxwell put it as follows:

> The 2nd law of thermodynamics has the same degree of truth as the statement that if you throw a tumblerful of water into the sea, you cannot get the same tumblerful out again.

In order to see such an unlikely configuration of the water molecules, we would have to wait for a time much longer than the present age of the universe!

Radioactivity and the birth of nuclear physics

Discoveries made in the last five years of the nineteenth century laid the foundations for our modern understanding of atomic and nuclear physics. The fascination of nineteenth century physicists with the strange effects produced by electricity in 'vacuum tubes' led to the discovery of X-rays, radioactivity and the electron in rapid succession. A 'vacuum tube' is a primitive version of the 'tube' that we now have in our present-day televisions. The early vacuum tubes consisted of a sealed glass cylinder with metal plates inserted at either end. When the plates were attached to a battery, the space between them glowed. The positively charged plate was called the 'anode' and the negative one the 'cathode'. By showing that an object placed in front of the cathode cast a shadow in the glow, the German physicist Johann Hittorf proved that the mysterious glow originated at the cathode. There was a great debate about the nature of these 'cathode rays' – similar to the great 'atoms versus thermodynamics' argument. Again, most of the German physicists shied away from a particle explanation and believed that cathode rays were a result of a wave motion. Most of the physicists working in England backed the idea that cathode 'rays' were made up of a stream of particles. In 1895, the French physicist Jean Perrin, who later performed some important experiments on Brownian motion, showed that cathode rays displayed some particle-like properties. Two years later, J. J. Thomson, director of the Cavendish Laboratory in Cambridge, conclusively demonstrated that cathode rays comprised negatively charged particles. He wrote that this explanation was contrary to 'the almost unanimous opinion of German physicists'.

Thomson's negatively charged particles were nearly 2000 times lighter than hydrogen, the lightest element, and were called 'electrons'. In fact, the name electron was not new; it had been introduced some years earlier by the Irish physicist George Johnstone Stoney to denote the minimum possible quantity of electric charge. In the eyes of the British physicists, the existence of the electron as an elementary particle implied the reality of atoms. Over the next ten years, Thomson went on to develop the 'plum pudding' model of the atom: a number of light, negative electrons embedded in a sphere of heavy, positively charged matter. This simple model gave rise to the first hope of understanding the differences and similarities between chemical elements in terms of stable patterns of electrons in atoms. The stage was now almost set for the entrance of Rutherford and Bohr to usher in the new age of atomic and nuclear physics. But there was one key discovery that had been missed by Thomson in his cathode ray experiments.

Wilhelm Roentgen was born in Germany, but spent most of his early life in his mother's native Holland. At the time he made his famous discovery, he was a professor at the University of Wurzberg in Germany. In November 1895, Roentgen was doing some experiments with a cathode ray tube covered entirely with black cardboard. The room was dark, and, to his surprise, he noticed that a test screen across the room was 'fluorescing'. Fluorescent

Figure 6.6 Sir Joseph John (J. J.) Thomson (1856–1940) was a bookseller's son who originally intended to become an engineer. He could not afford the charge made to become an apprentice engineer at Owens College (later Manchester University) and instead studied mathematics, physics and chemistry. Thomson won a scholarship to Cambridge, where he eventually became Cavendish Professor, developing the Cavendish Laboratory into one of the foremost research centres in the world. He would now be called an 'experimental physicist', but in reality he was rather clumsy, and his most important work was performed by assistants. He won the 1906 Nobel prize for physics, and seven of his research assistants were also awarded the Nobel prize. [Courtesy University of Cambridge, Cavendish Laboratory.]

Figure 6.7 Wilhelm Konrad Roentgen (1845–1923) was expelled as a student from Utrecht Technical School for laughing at a caricature of a teacher drawn by a fellow student and then being unwilling to divulge the name of the artist. This incident almost ended his academic career. In 1895, at the age of fifty, when he observed the X-ray projection of the bones in his hand for the first time, he was fearful that his colleagues might consider him a charlatan. Roentgen was not the first to make an X-ray photograph, but he was the first to realize its significance. In 1901, he was awarded the first Nobel prize for physics. [Courtesy AIP Emilio Segrè Visual Archives, Landé Collection.]

screens may be used to detect the arrival of fast-moving particles. They consist of a sheet of material coated with a 'phosphor', which gives off light when struck by particles. If the light stops immediately after the beam of particles is turned off, this effect is known as 'fluorescence': if the light persists as an 'afterglow', the effect is called 'phosphorescence'. In modern cathode ray tubes, zinc sulphide is commonly used as a phosphor: this gives off a blue light and has no afterglow. What was causing Roentgen's screen to give off light? Because the tube was covered, no light or cathode rays could reach the screen. Roentgen experimented by placing a number of objects in between the screen and the tube. To his great surprise, when he held his hand in the path of the 'rays', he found that he could see the bones of his hand projected on the phosphor screen! His subsequent paper announcing his discovery of 'X-rays' was ready in January of 1896, and it caused a sensation. Since Roentgen had included an X-ray photograph of his wife's hand with the paper, physicists were forced to accept his results. One did not need to be a physicist to see the implications of the discovery for medicine. In 1896, there were over 1000 papers on the subject of X-rays. In Cambridge, their discovery prompted Thomson to set the young Ernest Rutherford, who had newly arrived from New Zealand, to work on them. Before Rutherford had got into his stride, however, another discovery stole the scene.

There are many stories of famous discoveries by 'accident' in science. One of the best known is the discovery of 'Becquerel rays', which we now understand to be the result of radioactivity. Henri Becquerel was a professor in Paris, and was an expert on fluorescence and phosphorescence. On hearing about Roentgen's X-rays, he decided to find out if fluorescent materials gave off X-rays when exposed to sunlight. He put a layer of a uranium compound on top of a photographic plate sealed in black paper and left it in the sun for a few hours. When the photograph was developed, he saw an outline of the phosphor on the plate. Becquerel thought that this proved that sunlight released X-rays from the phosphor, and he duly reported this to the French Academy of Science on February 24th 1896. Only a week later, the weather had changed, and Becquerel soon knew he was wrong. When he tried to repeat the experiment on February 26th and 27th there was little or no sunshine, so Becquerel put everything away in a drawer of his desk. When the photographs were developed on March 1st, he expected to see only a very weak 'silhouette' of the uranium compound. As he reported: 'On the contrary, the silhouettes appeared with great intensity. I thought at once that the action might be able to go on in the dark.' Eight days later, Becquerel had found that the 'rays' given off by the uranium caused gases to conduct electricity. We now understand that the radiation from the uranium ionizes the gas by 'knocking' electrons out of the gas atoms. At the time of Becquerel's discoveries, no one knew about electrons, so these results were all very mysterious. This property of Becquerel rays to ionize gas made it much easier to measure the 'activity' of different samples by using electroscopes. These instruments were similar to those used to study static electricity.

Figure 6.8 One of the first X-ray pictures: the bones in the hand of Roentgen's wife. [Science Photo Library.]

Figure 6.9 Henri Becquerel (1852–1908), like his father and grandfather before him, was interested in fluorescence. He shared the 1903 Nobel prize with the Curies for the discovery of the radiation emitted by uranium salts. [The Mansell Collection.]

At this point in the history of physics, another heroic figure enters the frame. Two Polish sisters, Bronya and Marie Sklodowska, were determined to survive the loss of their father's fortune and complete their education. First, Bronya went to Paris to study medicine, supported in part by Marie working as a governess back in Poland. Then it was Bronya's turn to support Marie. Marie wanted to study physics, and she arrived in Paris in 1891 with the equivalent of about twenty dollars. She met the French physicist Pierre Curie three years later, and they were married in July of 1895. They spent their honeymoon cycling through the French countryside. After the birth of her first daughter, Irene, who later won a Nobel Prize in her own right, Marie Curie took her husband's advice for a subject for her doctoral research. She decided to investigate the new discoveries made by Becquerel, and she examined 'all the elements then known' for similar effects. Marie Curie found that only thorium emitted rays 'similar to those of uranium' and introduced the word 'radioactivity' to describe these effects. But she did not stop there. She decided to examine all the naturally occurring ores of these two radioactive elements, uranium and thorium. To her surprise, she found that, in some cases, the effects of radioactivity were three or four times greater than could be explained by the uranium or thorium content of the samples. Marie Curie correctly concluded that these samples of ore must contain a new radioactive element. In fact, the unknown substance was many million times more radioactive per unit weight than uranium, and it was present in the ore samples only at the level of about one part per million. Not knowing how much work would be involved, Marie, with her husband Pierre, then undertook the back-breaking task of grinding down the ore and treating it chemically to concentrate the most radioactive products. In July 1898 they announced the discovery of a new radioactive element, which Marie Curie called 'polonium' after her homeland.

Polonium, element 84 on the Periodic Table, resembles bismuth, element 83, except that it is a softer metal and emits so much radiation that ionized air around a sample glows with an eerie blue light. Polonium played a vital part both in the discovery of the neutron and as the initiator of the explosive chain reaction in the first atomic bomb. The Curies also showed that polonium 'decays' spontaneously and that it takes a specific time for a sample to reduce its radioactivity to half its initial size. This time is called the radioactive 'half-life', and it proves to be a characteristic signal for each radioactive element. Measuring the half-lives of the different radioactive signals enables physicists to disentangle the many different products that occur in the complicated radioactive decay chains of elements. The Curies went on to isolate their most famous discovery, the new radioactive element, radium, a few months later. It is still incredible that these successes were achieved without a properly equipped laboratory, wet in winter and hot in summer. In these inadequate surroundings, Marie and Pierre dissolved over 100 kilograms of ore by hand, stirring in chemical solvents without the safety of extractor fans to remove the dangerous fumes. Since no one had any idea of the dangers of radiation, it is not surprising that they both began to suffer from strange and hard to diagnose diseases. Becquerel was one of the first

Figure 6.10 A picture of Pierre (1856–1906) and Marie (1867–1934) Curie portrayed working in their laboratory, which appeared on the front cover of the magazine *Le Petit Parisien*. The draughts in their inadequate laboratory actually helped to prevent dangerous levels of airborne radiation building up. No precautions were taken against radioactivity, as the harmful effects were unknown. Marie Curie's notebooks, and even her cookery book, were later discovered to be highly radioactive.

people to be burned by radiation: he had put a sample of radium given to him by the Curies into his waistcoat pocket. The Curies also suffered from radiation burns. In the summer of 1903, Rutherford visited them in Paris and recorded that:

> we retired about 11 o'clock in the garden, where Professor Curie brought out a tube coated in part with zinc sulphide and containing a large quantity of radium in solution. The luminosity was brilliant in the darkness and it was a splendid finale to an unforgettable evening.

Rutherford was also able to see Pierre Curie's hands 'in a very inflamed and painful state due to exposure to radium rays'. Given such a cavalier attitude to handling radioactive materials, it was not surprising that Marie Curie's cookery books were found to be still radioactive fifty years after her death.

In 1900, the Curies prepared a report for an international physics conference in Paris on their discoveries. They ended their report by highlighting the two outstanding unsolved problems: the nature of the radiation and the source of its energy. Polonium decayed by emitting very energetic particles, but where did this energy come from? The answer lies with Einstein's mass–energy relation, but in 1900 this was not known. Nonetheless, the Curies suspected that some of their discoveries might be dangerous in the wrong hands. They were awarded the Nobel Prize in 1903, and in his acceptance speech Pierre Curie ended by reflecting on Alfred Nobel's discovery of dynamite:

> It is conceivable that radium in criminal hands may become very dangerous, and here one may ask whether it is advantageous for man to uncover natural secrets, whether he is ready to profit from it or whether this knowledge will not be detrimental to him. The example of Nobel's discoveries is characteristic; explosives of great power have allowed man to do some admirable works. They are also a terrible means of destruction in the hands of the great criminals who lead nations to war. I am among those who believe, with Nobel, that mankind will derive more good than evil from new discoveries.

The atom and the nucleus: Ernest Rutherford, Niels Bohr and Manchester

Our modern understanding of the atomic theory of matter and of the Periodic Table of the elements began in Manchester, in the north of England. It was here that the New Zealand-born physicist Ernest Rutherford showed that J. J. Thomson's 'plum pudding' model of the atom was wrong. Rutherford had been puzzling for more than a year on how to make sense of some experimental results obtained in Manchester by two of his protégés, Hans Geiger, of 'Geiger counter' fame, and an undergraduate student named Ernest Marsden. In 1911, Rutherford finally realized that almost all the mass of the atom and all its positive charge must be concentrated in a tiny part of the volume occupied by the atom, and consequently that the atom is mostly empty space. Phillipp Lenard, a German physicist, Nobel Prize winner and Nazi, had described this state of affairs in very graphic terms. He said that 'the space occupied by a cubic metre of solid platinum was as empty as the space of stars beyond the earth'. Before we recount the details of Rutherford's discovery of the atomic nucleus and how this catalysed the young Niels Bohr's invention of quantum theory, we must step back a few years to Rutherford's arrival at the Cavendish Laboratory in Cambridge in September of 1895.

Rutherford was stimulated by the discoveries of Becquerel and the Curies, and in 1898 he began a systematic exploration of the radioactive elements uranium and thorium. He rapidly came to the conclusion that:

> There are present at least two distinct types of radiation – one that is very readily absorbed, which will be termed for convenience alpha radiation, and the other of a more penetrative character, which will be termed the beta radiation.

Figure 6.11 Sir Ernest Rutherford (Lord Rutherford of Nelson) (1871–1937), at the front of the picture. The Russian physicist, Peter Kapitza, who worked with Rutherford, wrote to his mother in Leningrad (St Petersburg) about his colleague as follows: 'The Professor is a deceptive character. They (the English) think he is a hearty colonial. Not so. He is a man of immense temperament. He is given to uncontrollable excitement. His moods fluctuate violently. It will need great vigilance if I am going to obtain, and keep, his high opinion.' [Courtesy University of Cambridge, Cavendish Laboratory.]

Later, a French scientist, Paul Villard, identified a third distinct type of radioactivity, which he then called gamma radiation, following Rutherford's naming scheme. We now know that alpha particles are positively charged helium nuclei; beta particles are energetic electrons; and gamma rays are high-energy X-rays. At the time, it was very confusing, and it was some years before the physical processes involved were understood. Several of Rutherford's major discoveries were made when he was at McGill University in Canada. His letter of recommendation from J. J. Thomson for his appointment as a professor at McGill stated

> I have never had a student with more enthusiasm or ability for original research than Mr Rutherford, and I am sure if elected, he would establish a distinguished school of physics at Montreal.

In Montreal, Rutherford was lucky to have an unusually generous and far-sighted chairman of the physics department, John Cox. After observing Rutherford at work for a few weeks, Cox volunteered to take Rutherford's classes and do his teaching so that Rutherford could get on with research. After making a series of important discoveries, Rutherford found he needed the help of a chemist in order to make further progress in disentangling the

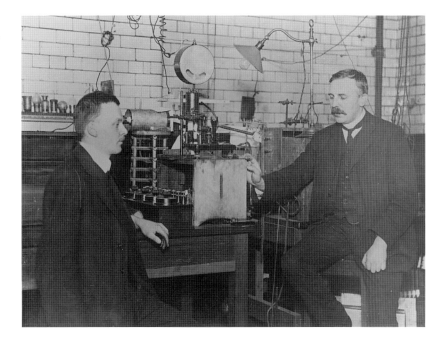

Figure 6.12 Hans Geiger (1882–1945) with Rutherford (at right) and their apparatus for counting alpha particles at Manchester University. [The University of Manchester.]

Figure 6.13 William Thomson (1824–1907) became professor of natural philosophy at Glasgow University at the age of twenty two, a post he occupied until his retirement at the age of seventy five. In 1866, he was knighted for his contribution to the laying of the first Atlantic telegraph cable, and in 1892 he was created Baron Kelvin. When Kelvin retired he enrolled as a research student. [The Annan Collection.]

complicated picture of radioactive processes he had uncovered. In the chemistry department at McGill, Rutherford found just the man he was looking for, an Englishman named Frederick Soddy. Together, Rutherford and Soddy were able to unravel the mysteries of radioactivity in a sequence of fundamental papers.

In 1900, Rutherford found that radioactive thorium gave off a radioactive gas. This finding suggested to Rutherford the idea that 'transmutation' of one element into another was occurring. Even Rutherford was reluctant to commit himself to this idea in print, since any 'transmutation of the elements' brought back memories of the discredited attempts of the alchemists to turn lead into gold. After talking with Rutherford about his results, Soddy tried to identify the gas given off by thorium. He found it was chemically inert. In Soddy's words, this

> conveyed the tremendous and inevitable conclusion that the element thorium was slowly and spontaneously transmuting itself into argon gas!

Using the different radioactive half-lives of the various radioactive elements as a signature to identify different stages in the decay chain, Rutherford and Soddy traced the way uranium, thorium and radium changed into different elements as they emitted alpha and beta particles.

In a paper written in 1903, Rutherford and Soddy made the first calculations of the energy released in radioactive decays. They concluded:

> The energy of radioactive change must therefore be at least twenty-thousand times, and may be a million times, as great as the energy of any molecular change.

The origin of this energy remained a mystery, but its existence enabled Rutherford to settle a long-running dispute between the physicist Lord Kelvin and the geologists. Kelvin had calculated the age of the Earth by estimating how long it would take the Earth to cool from a molten state to its present condition. Kelvin's estimate was much shorter than the age estimated by geologists from the geological evidence. Rutherford realized that the heat supplied by the decays of radioactive elements in the rocks of the Earth resolved the paradox. When Rutherford presented his theory during an open lecture, he was alarmed to see the intimidating Lord Kelvin in the audience. As Rutherford reached the critical point of his lecture, he was worried about how he could present his conclusion – that he agreed with the age estimated by the geologists – in a way that would not antagonize Kelvin. Rutherford resolved his problem with a master stroke:

> Then a sudden inspiration came and I said Lord Kelvin had limited the age of the earth, provided no new source was discovered. That prophetic utterance refers to what we are now considering tonight, radium! Behold, the old boy beamed upon me.

Rutherford left McGill in 1907 and moved to Manchester in the
north of England. There he set out to examine the structure of the atom by
firing beams of alpha particles at thin metal foil targets. With his assistant
Hans Geiger, he developed an instrument that made a click when an alpha
particle arrived: this was the fore-runner of the famous Geiger counter.
Geiger and the undergraduate Ernest Marsden were having some difficulty
with stray particles arriving at their detector, which were being confused with
the alpha particles scattered from the target. At this critical point, Rutherford
wandered into the laboratory and, after some discussion about the experiment, made the seemingly nonsensical suggestion that Marsden should look
and see if any alpha particles were being reflected back from the target. This
was a remarkable suggestion because the 'plum pudding' model of the atom
predicted that the alpha particles should go almost straight through the metal
foil with only small deflections – not bounce back towards the source.
Marsden performed the experiment very carefully and was surprised to see
alpha particles back-scattered from the target. Rutherford, despite his original suggestion, was genuinely astonished when Marsden told him the results:

> It was almost as incredible as if you fired a 15-inch shell at a piece of tissue
> paper and it came back and hit you.

After pondering on these results for over a year, Rutherford realized that they
implied that almost all the mass of an atom must be concentrated in a tiny
central region of the space occupied by the atom. Rutherford called this
central core the 'atomic nucleus', and he decided to make public his idea at
a meeting of a local organization called the Manchester Literary and
Philosophical Society. His lecture was second on the agenda, after the
announcement of the discovery of a rare snake in a consignment of fruit that
had been imported at Manchester. The date was March 7th, 1911.

The discovery of the nucleus by Rutherford led to the familiar
model of the atom as a miniature planetary system: the nucleus as the Sun
with the electrons, like the planets, in orbit around it. The problem for physicists was that, although there must be some grain of truth in such a 'planetary' model of the atom, according to the laws of classical physics such an
atom would be unstable. Unlike the solar system, the forces in the atom are
electrical, not gravitational. Maxwell's laws then predict that the charged
electrons, when accelerated in a circular orbit, lose energy by radiation and
spiral into the nucleus. The time estimated for such a catastrophic atomic
collapse is about one-million-millionth of a second! At this moment of crisis
for physics Niels Bohr arrived in Manchester.

Niels Bohr had defended his doctoral thesis in Copenhagen in May
of 1911 in an examination 'whose short duration is a record'. This was not
surprising since one of the examiners had said that hardly anyone in
Denmark was competent to judge Bohr's work. After a summer break, Bohr
went to study in Cambridge on a fellowship from the Carlsberg Foundation.
Although he found many aspects of Cambridge life very enjoyable, Bohr said
later: 'It takes half a year to know an Englishman.'

When Rutherford came to speak at the annual Cavendish dinner, Bohr was immediately impressed by the enthusiastic informality of the New Zealander. Later in life, Bohr remembered one of Rutherford's great qualities as having 'the patience to listen to every young man when he felt he had an idea, however modest, on his mind.'

After spending only about six months in Cambridge, Bohr moved to Manchester to work with Rutherford at the end of March, 1912. Rutherford was the quintessential experimentalist. He literally performed his experimental wizardry using sealing wax: red sealing wax from the Bank of England was used to make his glass tubes airtight. In contrast, Bohr was the model theoretician. Despite the fact that Rutherford normally distrusted theoreticians, he and Bohr got on well from the outset. When asked about this, Rutherford replied: 'Bohr's different. He's a football player!' Bohr *was* different – by mid-June he had the first insight about how to stabilize Rutherford's nuclear atom. He wrote to his brother: 'It could be that I've perhaps found out a little bit about the structure of atoms.' It took Bohr another nine months before he was able to send Rutherford the finished paper entitled 'On the constitution of atoms and molecules'.

The final push that enabled Bohr to finish his paper was from an old student friend, Hans Hansen. Hansen asked Bohr how his model accounted for the spectral lines of hydrogen. Bohr had no idea – so Hansen advised him to look up Balmer's formula for the wavelengths of the lines observed in the hydrogen spectrum. Bohr said later: 'As soon as I saw Balmer's formula the whole thing was immediately clear to me.' How did Bohr solve the problem of stability of the electron orbits? He didn't. He proposed instead that, for electron orbits that satisfied a new 'quantum condition', the normal laws of classical physics did not apply. To understand where this quantum condition originated, we must return again to another of Einstein's contributions to

Figure 6.14 The Bohrs and the Rutherfords at Cambridge. Niels Bohr (1885–1962), more than any other scientist, was responsible for leading physicists through the difficult transition years from his own early quantum ideas about atoms until the 'final' formulation of quantum theory. He founded the first international institute for theoretical physics, and before the Second World War all the leading physicists came to visit him in Copenhagen. The conventional way of reconciling the predictions of quantum theory with experiment is known as the Copenhagen interpretation. [The Niels Bohr Archive, Copenhagen. Photograph by Marcus Oliphant.]

twentieth century physics, and indeed the one for which he was officially awarded the Nobel Prize for Physics.

Quanta had been reluctantly introduced into physics by Max Planck to explain a peculiarity in the spectrum of light radiated from a hot 'black body'. Planck then spent many years unsuccessfully trying to find ways of evading his own conclusions. His idea had been taken up by Einstein in one of his 1905 papers to explain the 'photo-electric' effect, as discussed in the previous chapter. Einstein had shown that, in its interaction with the electrons of the metal, light should be regarded as a stream of particles, now called photons. The 'quantum', or packet, of energy possessed by the photons was found to be proportional to the frequency of the light; the constant of proportionality is called Planck's constant. This constant of Nature is so small that the 'quantized' properties of light are not easy to see in everyday life.

Bohr's new quantum theory of the atom was an ingenious mixture of classical and quantum physics. His allowed electron orbits were just those whose 'angular momentum' – the momentum of the electron multiplied by the radius of its orbit – was a multiple of Planck's constant. These allowed electron orbits had definite energies – a very different situation from a classical picture, for which any energy is allowed. Bohr then made contact with Balmer's famous formula and Einstein's relation of the energy of the photon to the frequency of the light. He suggested that the electron could be 'excited' into one of the higher energy levels. When the electron 'jumped' back to a lower energy level, a photon with an energy equal to the difference in energy of the two levels was emitted. Bohr's predictions were stunningly successful. Not only did his model of the atom correctly predict the frequencies of the spectral lines of hydrogen in the visible region of the electromagnetic spectrum, but it also predicted new spectral series in both the ultraviolet and the infrared parts of the spectrum.

In addition to his model of the atom, Bohr was also first to gain the first glimmer of an understanding of the Periodic Table of the elements. In 1922, the same year he later received the Nobel Prize for his atomic model, Bohr outlined the now traditional picture of atoms comprising a series of orbital 'shells' – like the nesting of a set of Russian dolls – with each shell only capable of accommodating a certain number of electrons. It is remarkable that Bohr was able to arrive at such a picture without the benefit of the full apparatus of quantum mechanics, and without knowledge of either 'electron spin' or of the Pauli Exclusion Principle. Bohr also predicted that the, as yet undiscovered, element with atomic number 72 – a positive charge of 72 on its nucleus – would not be a so-called 'rare earth' element as expected by the chemists, but an element like zirconium with chemical valency 4. George de Hevesy, a Hungarian radiochemist who had taught Bohr some chemistry when they were both in Manchester, was now working at Bohr's new physics institute in Copenhagen. He and a young Dutchman called Dirk Coster set to work to look for this new element in zircon-bearing minerals. On the night before Bohr's Nobel lecture, they called Bohr in Stockholm to tell him he was right. They named the new element hafnium, after the old Roman name for

Copenhagen, and Bohr was able to announce its discovery in his lecture the next day.

Perhaps the best tribute to Bohr was given by one of the founders of the new quantum mechanics, Werner Heisenberg. Heisenberg was a student in Munich when he and his colleagues learned of Bohr's predictions about the structure of atoms with ten, twenty or thirty electrons in different orbits. They could not understand how Bohr had obtained these results. Astronomers had not been able to give an exact solution to even the classic three-body problem involving the Sun, Earth and Moon. Bohr must be an intimidatingly clever mathematician to solve problems with up to thirty electrons. It was only when Heisenberg met Bohr some time later that he learnt how Bohr worked. Of this first meeting, Heisenberg said:

> The first, for me, quite shocking experience was that Bohr had calculated nothing. He had just guessed his results.

When Heisenberg then asked him if he thought anyone could calculate these results using classical Newtonian mechanics, Bohr gave the following reply:

> We are now in a new field of physics, in which we know that the old concepts probably don't work. We see that they don't work, because otherwise atoms wouldn't be stable. On the other hand when we want to speak about atoms, we must use words and these words can only be taken from old concepts, from the old language. Therefore we are in a hopeless dilemma, we are like sailors coming to a very far away country. They don't know the country and they see people whose language they have never heard, so they don't know how to communicate. Therefore, so far as the classical concepts work, that is, so far as we can speak about the motion of electrons, about their velocity, about their energy etc., I think my pictures are correct, but nobody knows how far such a language goes.

Heisenberg later credited Bohr with changing his whole attitude to physics. Heisenberg was one of the pioneers of the new quantum mechanics, which, in the space of the next few years, replaced Bohr's 'old' quantum theory.

Relativity, quantum mechanics and electron spin

Three definitive papers were published in 1925 and 1926 that set down the principles of the new quantum mechanics: they were by the German physicist Werner Heisenberg, together with Max Born and Pascual Jordan; by the Austrian physicist Erwin Schroedinger; and by a remarkable English physicist named Paul Dirac. The new quantum theory was able to reproduce Bohr's energy levels for the hydrogen atom as well as to explain problems that Bohr's model had been unable to answer. Dirac then showed how to include the electromagnetic field in quantum mechanics in a consistent way. This enabled physicists to calculate, for the first time, the relative probabilities of different 'quantum jumps' for atomic electrons. The problem of two-electron atoms, such as helium, for which Bohr's model failed badly, was also solved by the new quantum mechanics. So, how does all this relate to Einstein's relativity? Both the 'old' quantum theory and the 'new' quantum mechanics were based on the assumption of non-relativistic motion for the atomic electrons. In other words, the electron's velocity was assumed to be much less than the speed of light. In fact, electron velocities were an appreciable fraction of light speed, and one ought to be able to calculate 'relativistic corrections' to the energy levels of the hydrogen atom calculated assuming non-relativistic motion. This had been done for the Bohr model by Heisenberg's professor in Munich, Arnold Sommerfeld. Sommerfeld's relativistic corrections had predicted that some spectral lines should, in fact, comprise two lines close together. This so-called 'fine structure' was soon confirmed experimentally. It was clearly necessary for the new quantum mechanics to be able to reproduce this success.

There were more surprises and problems in store. Although Sommerfeld's model worked well for hydrogen, it did not fit the data for the alkali atoms. The alkali atoms have a lone electron orbiting a closed shell. In Bohr's model the expectation is that this electron should be moving more slowly than the one in the hydrogen atom. Accordingly, Sommerfeld's theory predicted a smaller relativistic correction to the energy levels and a smaller fine-structure splitting of the spectral lines. Experiment shows the opposite. In November, 1925, two Dutch graduate students working in Leyden, George Uhlenbeck and Sam Goudsmit, suggested that besides having some 'orbital' angular momentum due to its motion round the nucleus, the electron possessed what they called a 'spin' angular momentum. This can be described as being rather like the Earth spinning on its axis as it goes round the Sun. Uhlenbeck and Goudsmit's professor, Paul Ehrenfest, suggested that they write up their suggestion as a letter for a journal and 'then we will ask Lorentz'. This they duly did, and, after a week or so, Lorentz came back to them with some calculations, showing that, according to classical physics, the whole idea made no sense. Uhlenbeck and Goudsmit then rushed off to Ehrenfest to tell him not to submit their paper for publication. To their dismay – and later relief – Ehrenfest replied:

> I have already sent your letter in long ago; you are both young enough to allow yourselves some foolishness!

In fact, their insight had revealed a new, non-classical property of the electron.

What has electron spin got to do with relativity and the fine structure? The new quantum mechanics predicted that there should be a contribution to the energy of the electron arising from an interaction between the spin of the electron and its orbital motion. Roughly speaking, the motion of the electron in its orbit can be regarded as a loop of electric current. Such a current will give rise to a magnetic field. Also according to classical physics, because of the electron's spin angular momentum, one expects the electron to behave like a little bar magnet. The interaction of this 'spin' magnet with the 'orbital' magnetic field, called the 'spin–orbit coupling', gives a calculable contribution to the electron energy. This shows itself in the fine-structure splitting of the spectral lines. In the calculation, there are two puzzling factors of 2 which have to do with relativity. The first one appears in the relation between the electron's spin angular momentum and the strength of the spin magnet produced. Classically, the constant of proportionality – the 'g-factor' – was expected to be unity. Experiments with electron beams in magnetic fields showed that the electron g-factor was, in fact, very nearly 2. The second factor of 2 was just as surprising. A student at Cambridge with Dirac, Llewellyn Thomas, had shown that there was a new effect of Einstein's relativity to take into account. He pointed out that people had been calculating the energy as if the electron were at rest with the nucleus swirling around it. Instead, what was needed to compare theory with experiment was the energy calculated with the nucleus at rest. Thomas used special relativity to show that this simple change of perspective resulted in a reduction of the energy shift by a factor of 2. Inclusion of such relativistic 'corrections' is standard practice in the calculations of physicists and chemists today.

Dirac and anti-particles

Paul Dirac was a strangely shy but brilliant physicist. His early work on quantum mechanics was completed whilst still a Ph.D. student at Cambridge. Dirac later married the sister of Nobel Prize winner Eugene Wigner. When speaking of the flamboyant American physicist, Richard Feynman, Wigner is reported to have said: 'He is a second Dirac, only this time human.' Dirac was not only famous for his research contributions, he also wrote up his early work on quantum mechanics in one of the first textbooks on the subject. In this now classic text entitled *The Principles of Quantum Mechanics*, Dirac adopts an abstract mathematical approach to the theory, which not everyone found easy – not even some of the other founders of quantum mechanics. The Russian translation included a warning that Dirac's book contained – 'a whole series of opinions, both explicit and implicit, which are totally incompatible with Dialectical Materialism'!

Dirac's early contributions to the development of quantum mechanics took place during the years from 1925 to 1928. These contributions alone would have been enough for him to be listed as one of the great physicists of this century, yet it is for his 1928 paper on the 'relativistic theory of the electron' that he is most famous.

Instead of beginning with the non-relativistic energy–momentum relation for a free particle, which relates the particle's energy to the square of its momentum, Dirac wanted to start from the relativistic relation, in which energy and momentum appear on the same footing. Several people, including Schroedinger, had tried starting from the relativistic relation involving the square of the energy and the square of the particle momentum. This approach leads to several problems, the most serious of which is that a

Figure 6.15 Richard Feynman (on the right) talking to Paul Adrien Maurice Dirac (1902–1984) in Warsaw in 1962. The conversation is likely to have been either one-sided or rather sparse as Dirac was famous for being taciturn. Dirac was, nevertheless, Feynman's physics hero. [Courtesy of the Archives, California Institute of Technology.]

quantity normally interpreted as a probability can be negative. Schroedinger became discouraged, and he returned to the non-relativistic relation to derive his famous equation. Characteristically, Dirac adopted a frontal attack on the problem. He started by writing down an equation in which the energy and momentum appeared on an equal footing. To his surprise, he found that his equation only made sense if it was interpreted as a 'matrix' equation. Matrices are well-known mathematical objects consisting of an array of numbers, with a well-defined set of rules for multiplication, addition and so on. In particular, given two matrices A and B, it is possible for A times B to be different from B times A. It was this same property that had first drawn Heisenberg's attention to matrices in his approach to quantum mechanics. Dirac's matrices consisted of 4×4 arrays of complex numbers. What was very puzzling was that the quantity representing the electron was a 'column matrix' with four entries. Physicists had already realized that they needed column matrices with two components to accommodate the electron spin. What was the meaning of Dirac's two extra components?

Dirac found that his equation eliminated the negative probabilities, but there was another problem: Dirac's electron could have negative as well as positive energies! Ignoring this for the time being, Dirac went on to apply his theory to the hydrogen atom. As if by magic, his equation correctly predicted the observed fine structure of the energy levels. In particular, it correctly predicted that the mysterious 'non-classical' *g*-factor of the electron should be 2, as found by experiment. Dirac's energy level formula for the fine structure turned out to be identical to Sommerfeld's formula derived from the old quantum theory. This is despite the fact that the physics of Sommerfeld's derivation was now seen to be completely wrong. This famous red herring has been described by the physicist John van Vleck as 'perhaps the most remarkable numerical coincidence in the history of physics'.

What is to be made of Dirac's negative energy levels? What stops the electron losing energy by jumping from a positive energy level to one of the negative energy levels? According to Wolfgang Pauli's 'Exclusion Principle', no two electrons can occupy the same energy level. Dirac therefore made the astonishing suggestion that all these negative energy levels were normally full. According to Dirac, a box that looked empty in fact contained an infinite number of electrons filling these negative energy states. Although this sounds absurd – an 'empty' box with infinite negative energy and charge – it is not so strange when one considers that one only measures energy and charge relative to this 'empty' box state. Dirac considered the effect of one of these negative energy states being empty – a 'hole'. Comparing this situation with Dirac's 'empty' box, we are now missing negative energy and negative charge. The empty box with a hole therefore looks like a box containing a particle of positive energy and charge! At the time, only two elementary particles were known to exist: the negatively charged electron and the positively charged proton, some 2000 times heavier than the electron. At first, Dirac tried to identify his positively charged 'hole' particle with the proton, even though his equation predicted that this new particle should have the same mass as the electron. He hoped some unspecified 'interactions' would later account for the difference in mass. But there was another serious difficulty for this proton interpretation. If we consider a box containing a positive energy electron as well as a hole, there is nothing to stop the electron falling into the hole's vacant energy level and being 'annihilated' in a flash of light energy. When Dirac calculated how long a hydrogen atom made up of a proton 'hole' and an electron could last before the two particles annihilated each other, the answer came out as a fraction of a second. Small wonder that Dirac's theory was received with scepticism around the world. Pauli jokingly suggested the 'Second Pauli Principle': 'Whenever a physicist proposes a theory, it should immediately become applicable to himself; therefore, Dirac should annihilate!' Even Dirac's friend Heisenberg wrote to Pauli saying: 'The saddest chapter of modern physics is and remains the Dirac theory.'

Dirac, however, had the last laugh. His equation proved to be a triumph of theoretical physics. By combining both quantum mechanics and relativity in a theory of great mathematical beauty, Dirac had discovered an entirely unexpected phenomenon – the world of anti-matter. In 1931, Dirac finally abandoned hope of identifying the positive energy holes with protons. In true Dirac fashion, he bravely concluded, in the total absence of any hint of experimental confirmation, that his equation predicted:

a new kind of particle, unknown to experimental physics, having the same mass and opposite charge to an electron.

Caltech physicist Carl Anderson discovered the anti-electron – or 'positron' as Anderson later called it – in August, 1932. At this time, Anderson was seemingly unaware of Dirac's prediction. Dirac's prediction was soon further substantiated when Patrick Blackett and Giuseppe Occhialini, working at the Cavendish Laboratory in Cambridge, observed the creation of electron–positron pairs in cosmic-ray showers – the converse of the annihilation process – in which energy is given to an empty box, exciting a negative energy electron to a positive energy level, leaving behind a positive energy hole, or positron, together with a positive energy electron.

Figure 6.16 Carl Anderson working with an electromagnet for use with his cloud chamber. [Courtesy of the California Institute of Technology.]

Figure 6.17 This cloud chamber photograph taken by Carl Anderson provided the first evidence for the positron. Note that the curvature of the track in the top half of the photograph is greater than at the bottom, which means that the particle in the top half has less energy. The particle must therefore have travelled from the bottom to the top since a charged particle will lose energy in passing through the lead plate that Anderson had fixed in the chamber. This trick allowed Anderson to be sure that the charge of the particle was positive, and he was also able to determine that its mass was the same as that of the electron. [Courtesy of the California Institute of Technology.]

With the possibility of pair creation and annihilation, relativistic quantum mechanics cannot properly be considered as a theory with a fixed number of particles. In addition, having to think of the empty state – the 'vacuum' – as containing an infinite set of negative energy levels was also very awkward. Later developments by the Nobel Prize trio of Richard Feynman, Julian Schwinger and Sin-itiro Tomonaga, and by Freeman Dyson, led to what is now the present state of the art: the theory known as 'quantum electro-dynamics', QED for short. This theory predicts elementary particle processes to an almost unbelievable accuracy. For example, the electron's g-factor is found to be not exactly 2 when even smaller 'higher order' corrections are calculated. QED predicts that there is an 'anomaly' a defined by the equation

$$g = 2(1 + a)$$

The latest published value for the electron anomaly is approximately

$$a = 11596522 \times 10^{-12}$$

Figure 6.18 The Bevatron at the Lawrence Berkeley Laboratory began accelerating protons in 1954 and is still working today. Protons are accelerated up to an energy such that their mass becomes more than six times their rest mass. The first antiprotons were produced using this machine. [Lawrence Berkeley National Laboratory.]

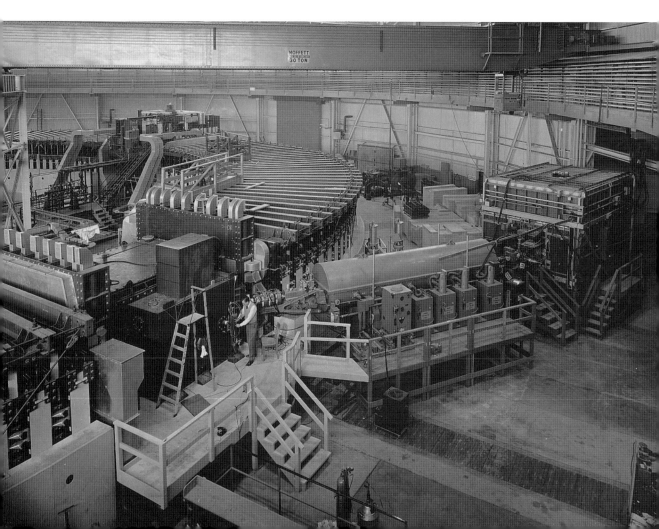

which is in remarkable agreement with the value predicted by QED,

$$a = 11596519 \times 10^{-12}$$

Such stunning numerical success, together with predictions for electron–positron pair creation and many other processes, make it clear that QED is the most successful and predictive theory ever invented by physicists. Nonetheless, there remain some distinguished physicists who still have some reservations. The basic worry can be understood in the context of the Dirac 'sea' of negative energy electrons. Dirac's empty state, the 'vacuum', has an infinite negative charge and an infinite negative energy. By considering only changes relative to this sea, one can ignore these troublesome infinities and make finite predictions about holes and positrons. In QED, there are a whole series of infinities besides this one. All these infinities can be 'tamed' in the sense that we can make definite finite predictions using a process known as 'renormalization'. Feynman was one of the inventors, yet he once said of renormalization:

> But no matter how clever the word, it is what I would call a dippy process! Having to resort to such hocus-pocus has prevented us from proving that the theory of quantum electrodynamics is mathematically self-consistent.

Feynman always suspected that renormalization is not mathematically legitimate. Dirac, too, remained a doubter to the end of his life. In one of his last public lectures, Dirac said:

> …we should no longer have to make use of such illogical processes as infinite renormalization. This is quite nonsense physically, and I have always been opposed to it. In spite of its successes, one should be prepared to abandon it completely and look on all the successes that have been obtained by using the usual forms of quantum electrodynamics with the infinities removed by artificial processes as just accidents when they give the right answers, in the same way as the successes of Bohr's theory are considered merely as accidents when they turn out to be correct.

The world of anti-matter does not only apply to the electron. Once Dirac had been forced to abandon the idea that his equation described both the two particles known to experiment, i.e. the electron and the proton, he followed through the consequences inexorably. If there were an anti-electron, Dirac argued there should also be an anti-proton. It was not until 1955 that particle accelerator technology advanced sufficiently for there to be enough energy available in a collision for a proton and anti-proton pair to be produced. Owen Chamberlain, Emilio Segrè, Clyde Wiegand and Thomas Ypsilantis then found the elusive anti-proton using the Bevatron accelerator in Berkeley, California.

In the conclusion of his 1933 Nobel Prize lecture, Dirac put forward an even more speculative idea:

We must regard it rather an accident that the Earth (and, presumably, the whole solar system), contains a preponderance of negative electrons and positive protons. It is quite probable that for some of the stars it is the other way about, these stars being built up mainly of positrons and negative protons. In fact there may be half the stars of each kind. The two kinds of stars would both show exactly the same spectra, and there would be no way of distinguishing them by present astronomical methods.

Nowadays, astrophysicists believe such speculation unlikely because of the absence of radiation from annihilations that would arise from matter–anti-matter collisions. It is now thought possible that the observed asymmetry between the existence of matter and anti-matter may be explicable in terms of a small difference in the way particles and anti-particles interact.

Aside from such speculations on the origin of matter, there are more pragmatic applications of Dirac's discovery of anti-matter. Besides radioactive decays emitting electrons, it is now known that there are many nuclei that are positron emitters. Positrons soon find another electron with which to mate, and this annihilation process results in a very characteristic burst of high-energy photons. This is the basis for the 'positron emission tomograph', or PET, in which positron-emitting radioactive tracer atoms are injected into human tissue: a PET scan then reveals the distribution of these atoms. Such a technique has been used to investigate the biochemical effects of anti-depressant drugs on brain function. All this seems a long way from Einstein and the Michelson and Morley experiment.

Figure 6.19 Positron emission tomography (PET) enables us to explore the activity of the brain in response to various stimuli. The patient is injected with a positron-emitting isotope which attaches itself to glucose in the blood. Since areas of increased brain activity are known to correspond to higher concentrations of glucose, these sites will display higher levels of radioactivity caused by the injected isotope. A positron emitted in the radioactive decay of the isotope is absorbed by the brain tissue and annihilates with an electron to form two gamma rays. These gamma rays have a characteristic energy and are emitted 'back to back'. Observation of these annihilation photons allows the source of radiation to be precisely located. [CERN photo.]

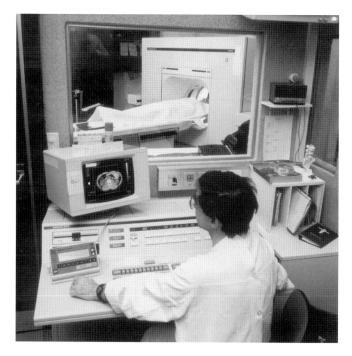

End-piece

Despite being a brilliant and far-sighted physicist, Dirac did not make idle conversation, and his responses to questions were often very surprising. A story told by the American physicist Alan Krisch will illustrate the point. Pauli and Dirac were riding on a train in the countryside. After an hour of silence, Pauli was becoming uncomfortable and was eagerly searching for an opening remark. Upon passing some sheep, he said to Dirac: 'It looks like the sheep have been freshly shorn.' After studying the sheep, Dirac replied: 'At least on this side.' There are many such 'Dirac' stories illustrating both his quirky originality and his unusual responses. Since we have talked about Dirac's 'hole theory' and his negative energies we end this chapter with the following two anecdotes.

Rudolf Peierls recounts the following story. A colleague in Cambridge, H. R. Hulme, was walking with Dirac when something rattled in his coat pocket. He apologized to Dirac for the noise, explaining that he had a bottle of pills and had taken some for his cold, so that the bottle was no longer full. After some time, Dirac replied: 'I suppose it makes the maximum noise when it is half full.'

The other story is probably apochryphal, but is certainly in character. The story centres round a puzzle. Five friends had been collecting coconuts, and, at the end of the day, they went to sleep round a camp-fire, promising to divide the coconuts equally between them in the morning. During the night, one of them woke up and decided to take his share then. He counted the coconuts, found that the number did not quite divide by five, but that if he gave one to the watching monkey, he could then take one-fifth of the number remaining. He tucked his share of the coconuts under his sleeping bag and went back to sleep. Then a second friend woke up. He repeated the process: he counted the coconuts, gave one to the monkey, took a fifth of the remainder and went back to sleep. Eventually all five of the friends have gone through these same actions. The puzzle is to figure out how many coconuts they started with. Supposedly, Dirac's answer was −4.

7   Little Boy and Fat Man: relativity in action

I do not believe that civilization will be wiped out in a war fought with the atomic bomb. Perhaps two-thirds of the people of the earth might be killed. But enough men capable of thinking, and enough books, would be left to start again, and civilization could be restored.

Albert Einstein, in *Atlantic Monthly*, 1945

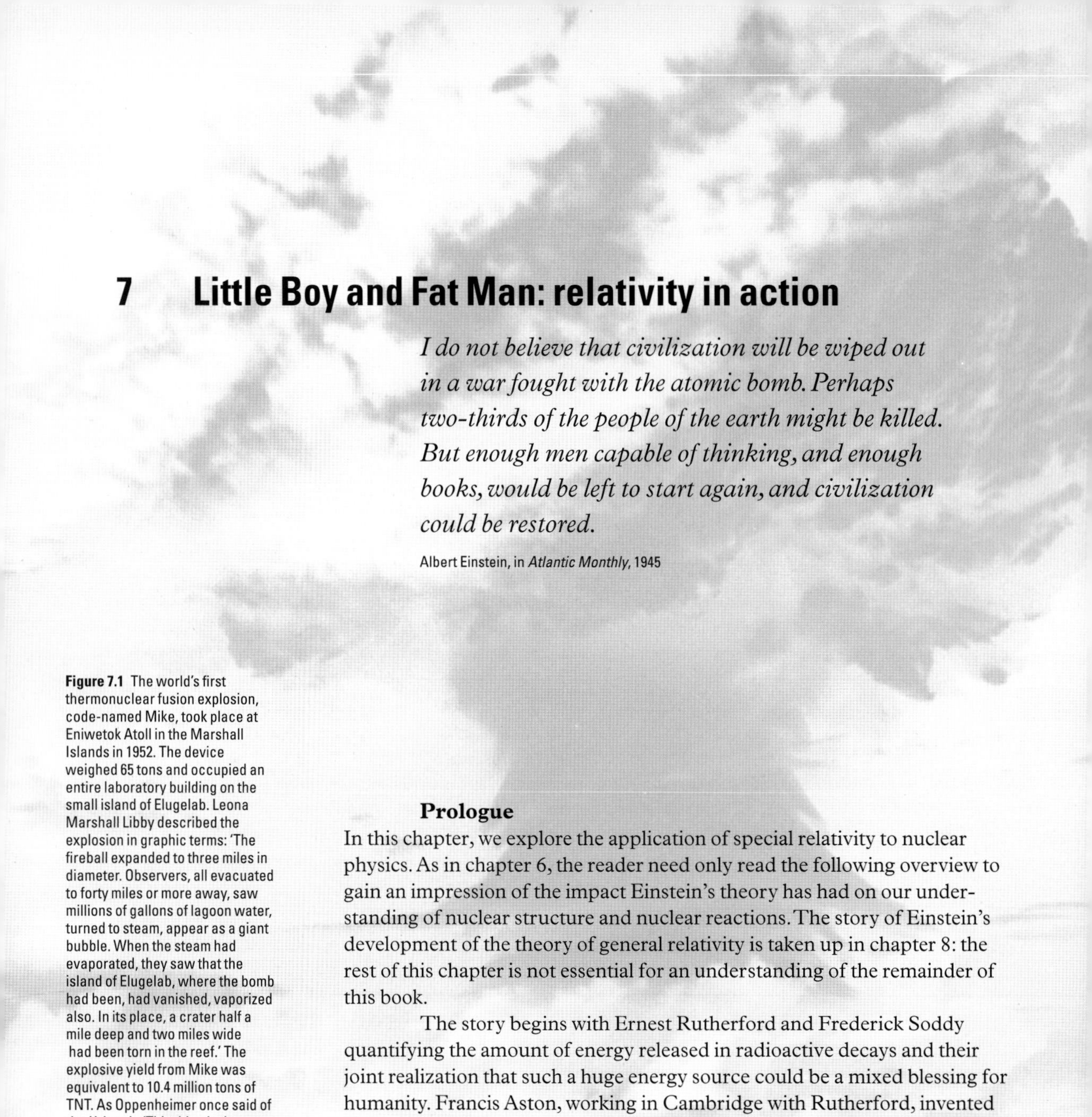

Figure 7.1 The world's first thermonuclear fusion explosion, code-named Mike, took place at Eniwetok Atoll in the Marshall Islands in 1952. The device weighed 65 tons and occupied an entire laboratory building on the small island of Elugelab. Leona Marshall Libby described the explosion in graphic terms: 'The fireball expanded to three miles in diameter. Observers, all evacuated to forty miles or more away, saw millions of gallons of lagoon water, turned to steam, appear as a giant bubble. When the steam had evaporated, they saw that the island of Elugelab, where the bomb had been, had vanished, vaporized also. In its place, a crater half a mile deep and two miles wide had been torn in the reef.' The explosive yield from Mike was equivalent to 10.4 million tons of TNT. As Oppenheimer once said of the H-bomb: 'This thing is the plague of Thebes'. [Los Alamos National Laboratory.]

Prologue

In this chapter, we explore the application of special relativity to nuclear physics. As in chapter 6, the reader need only read the following overview to gain an impression of the impact Einstein's theory has had on our understanding of nuclear structure and nuclear reactions. The story of Einstein's development of the theory of general relativity is taken up in chapter 8: the rest of this chapter is not essential for an understanding of the remainder of this book.

The story begins with Ernest Rutherford and Frederick Soddy quantifying the amount of energy released in radioactive decays and their joint realization that such a huge energy source could be a mixed blessing for humanity. Francis Aston, working in Cambridge with Rutherford, invented the 'mass spectrograph' and was able to separate different 'isotopes' of many elements. There was much confusion about the nature of 'isotopes' until James Chadwick discovered the neutron in 1932. Aston had also realized that the neutrons and protons when bound in the nucleus weigh less than in their free state. The difference arises from the nuclear binding energy, which results from the strong nuclear forces holding the nucleus together. This is a direct example of Einstein's mass–energy relation.

The next significant discovery in nuclear physics was the observation by Rutherford of the first artificial nuclear reaction – a genuine transmutation of elements. Enrico Fermi and his fellow Italian physicists then used the neutron to initiate nuclear reactions and produce a whole range of new unstable radioactive isotopes. With uranium, the heaviest known element, they made a mistake: they believed they had made a new 'trans-uranic' element. Four years later, on the eve of the Second World War, Otto Hahn and Fritz Strassmann, working in Berlin, identified barium among the uranium reaction products. Lise Meitner and Otto Frisch realized the significance of this result: under bombardment with a neutron, the uranium nucleus had broken apart into two lighter nuclei. They called this process 'nuclear fission'.

The energy released as a result of nuclear fission can be calculated using the mass–energy relation. As early as 1934, Leo Szilard had patented the idea of a chain reaction. With the discovery of fission, he realized that uranium held the key to developing a bomb. In 1939 he persuaded Einstein to write to President Roosevelt warning him of the possibility of nuclear weapons and urging that he authorize a full-scale programme to investigate their feasibility. The British had also recognized this possibility, and two refugees from Hitler's Germany, Otto Frisch and Rudolf Peierls, working in Birmingham wrote a famous memorandum demonstrating the feasibility and potential of such new weapons. Their report led to the setting up of the British atomic bomb programme, code-named 'Tube Alloys'. The same report was also influential in galvanizing the Americans into action and into setting up the US bomb programme – the Manhattan Project – and the Los Alamos Laboratory directed by Leslie R. Groves and J. Robert Oppenheimer. The practical results of this application of Einstein's mass–energy relation were the uranium and plutonium bombs dropped on Hiroshima and Nagasaki.

Science fact or science fiction?

It was in 1903 that Ernest Rutherford and his collaborator, Frederick Soddy, first quantified the huge amount of energy released in radioactive decay processes. Since this was before Einstein's famous mass–energy relation, the origin of this energy was a mystery. Both men were already aware that this energy was potentially dangerous. Rutherford was heard to say that 'some fool in a laboratory might blow up the universe unawares'. Soddy, in a lecture in 1904, speculated:

> It is probable that all heavy matter possesses – latent and bound up within the structure of the atom – a similar quantity of energy to that possessed by radium. If it could be tapped and controlled what an agent it would be in shaping the world's destiny! The man who put his hand on the lever by which a parsimonius Nature regulates so jealously the output of this store of energy would possess a weapon by which he could destroy the earth if he chose.

Soddy trusted 'Nature to guard her secret': the famous science fiction writer and novelist H. G. Wells was not so optimistic. In a little known novel, *The World Set Free*, published just before the First World War, Wells incorporated Soddy's speculation. This prophetic book was about both nuclear war and the nuclear stalemate that he foresaw would inevitably follow. In his story, world-wide control of nuclear weapons was only achieved at the cost of annihilation, by atomic bombs, of all the major cities in Europe. Only after such a catastrophe was it possible to concentrate the minds of politicians on control of nuclear weapons. Not surprisingly, since nuclear physics had barely begun, Wells got details wrong, but he also got a lot right. It is intriguing that his fictional bomb material 'carolinium', an artificial element 'most heavily stored with energy and the most dangerous to make and handle', has a direct parallel with plutonium. Plutonium was the artificial (man-made) element used in the 'Fat Man' atomic bomb which was dropped on Nagasaki in 1945. *The World Set Free* was not a successful book, and is now out of print. Although admitting that it was carelessly written, Wells thought it a good book, but agreed that 'nobody else has ever done so, and it was a failure'.

Wells was not a newcomer to controversy. In *The World Set Free*, Wells included a gentle reproach to his critics who had dismissed his earlier predictions of 'flying machines'. He recounted how his son came and spoke very seriously to him: 'I wish, Daddy, that you wouldn't write all this stuff about flying. The chaps rot me.' Wells was right in his insight about the possibility of flying, and also right about his vision of atomic bombs with a destructive power so vast that nobody could 'win' a nuclear war. But, despite

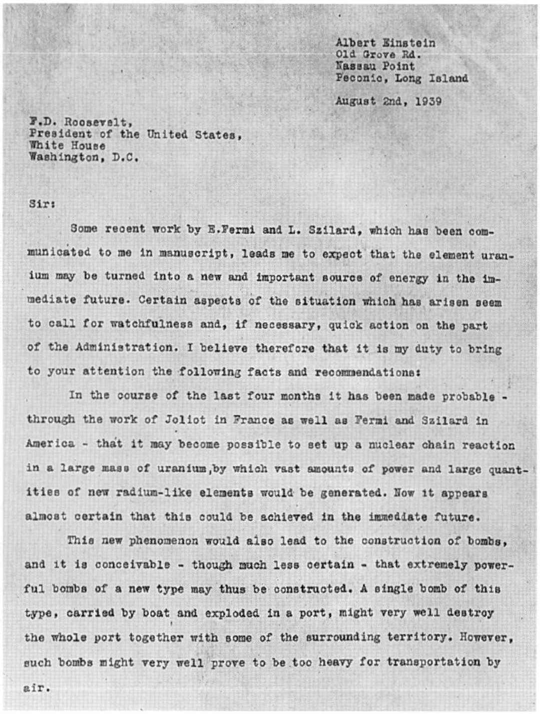

Albert Einstein
Old Grove Rd.
Nassau Point
Peconic, Long Island

August 2nd, 1939

F.D. Roosevelt,
President of the United States,
White House
Washington, D.C.

Sir:

Some recent work by E.Fermi and L. Szilard, which has been communicated to me in manuscript, leads me to expect that the element uranium may be turned into a new and important source of energy in the immediate future. Certain aspects of the situation which has arisen seem to call for watchfulness and, if necessary, quick action on the part of the Administration. I believe therefore that it is my duty to bring to your attention the following facts and recommendations:

In the course of the last four months it has been made probable - through the work of Joliot in France as well as Fermi and Szilard in America - that it may become possible to set up a nuclear chain reaction in a large mass of uranium, by which vast amounts of power and large quantities of new radium-like elements would be generated. Now it appears almost certain that this could be achieved in the immediate future.

This new phenomenon would also lead to the construction of bombs, and it is conceivable - though much less certain - that extremely powerful bombs of a new type may thus be constructed. A single bomb of this type, carried by boat and exploded in a port, might very well destroy the whole port together with some of the surrounding territory. However, such bombs might very well prove to be too heavy for transportation by air.

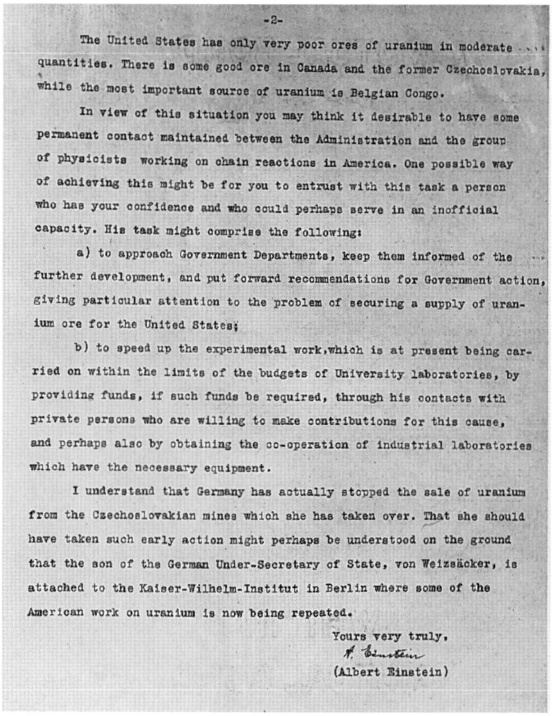

-2-

The United States has only very poor ores of uranium in moderate quantities. There is some good ore in Canada and the former Czechoslovakia, while the most important source of uranium is Belgian Congo.

In view of this situation you may think it desirable to have some permanent contact maintained between the Administration and the group of physicists working on chain reactions in America. One possible way of achieving this might be for you to entrust with this task a person who has your confidence and who could perhaps serve in an inofficial capacity. His task might comprise the following:

a) to approach Government Departments, keep them informed of the further development, and put forward recommendations for Government action, giving particular attention to the problem of securing a supply of uranium ore for the United States;

b) to speed up the experimental work, which is at present being carried on within the limits of the budgets of University laboratories, by providing funds, if such funds be required, through his contacts with private persons who are willing to make contributions for this cause, and perhaps also by obtaining the co-operation of industrial laboratories which have the necessary equipment.

I understand that Germany has actually stopped the sale of uranium from the Czechoslovakian mines which she has taken over. That she should have taken such early action might perhaps be understood on the ground that the son of the German Under-Secretary of State, von Weizsäcker, is attached to the Kaiser-Wilhelm-Institut in Berlin where some of the American work on uranium is now being repeated.

Yours very truly,
A. Einstein
(Albert Einstein)

Figure 7.2 Einstein's letter to President Roosevelt warning of the dangers of an atomic bomb. [Photograph courtesy of the Franklin D. Roosevelt Library.]

being a commercial failure, Wells' obscure novel did have a significant impact: the ideas in *The World Set Free* had made a great impression on the Hungarian emigré physicist Leo Szilard. Szilard had seen his Jewish friends and physicist colleagues driven out of Europe by the Nazis, and he was terrified by the idea that such a powerful weapon could fall into the hands of Hitler. Werner Heisenberg, one of the greatest physicists of this century, was still working in Nazi Germany, so, to Szilard, the development of an atomic bomb by the Nazis was a very real and alarming prospect. When Szilard reached America, he tried to alert the United States Government to the danger of an atomic bomb in the hands of Nazi Germany. After many months of frustration, trying unsuccessfully to rouse US officialdom to action, Szilard tried a less conventional approach. Before the war, Szilard had developed a patent for a novel type of refrigerator – it used a magnetic pump with no moving parts or seals that could break. The German company AEG built a working prototype of this refrigerator, but eventually decided not to market it. His co-inventor on the patent was Einstein. In despair at ever getting the US establishment to listen, Szilard turned to his old friend, also a refugee, then living near New York on Long Island. Szilard urged Einstein to write to President Roosevelt. Einstein's famous letter, drafted by Szilard, explained in simple terms the awesome destructive potential of nuclear weapons – able to destroy 'a whole port together with some of the surrounding territory' – and implored Roosevelt to initiate a crash programme to develop atomic weapons before Nazi Germany. So it was that H. G. Wells' 'failure' of a book played a part in this momentous piece of history. There is an ironic postscript to this story. Although Szilard had played a key role in initiating the development of atomic weapons, at the end of the war in Europe, after the defeat of Germany, he was one of the scientists most adamantly opposed to dropping the bombs on Japan.

The key to the nucleus

Today, we learn at an early age that the atomic nucleus contains protons and neutrons, and it is difficult to imagine the confusion that arose in the early days of nuclear physics.

The neutron is a neutral particle, slightly heavier than the proton. Each element is characterized by its atomic number – the number of protons in the nucleus – which is equal to the number of 'orbiting' electrons. The chemical properties of the elements are determined by these electrons. Each nucleus can contain a number of neutrons as well as protons. Any given element usually occurs with a particular number of neutrons combined with the protons to form its nucleus, but the same element can also be found in other less common forms with the nucleus containing different numbers of neutrons. For example, hydrogen usually occurs with a single proton as the nucleus, but about fifteen of every 100 000 hydrogen atoms is in the form of 'heavy hydrogen' or deuterium. Deuterium has a proton and a neutron bound together in its nucleus, and plays a very important role in nuclear physics and stellar evolution. Deuterium is said to be an 'isotope' of hydrogen, which means that it has the same chemical properties as hydrogen, and only differs in the mass of its nucleus. Another example is provided by uranium. The most common isotope of uranium is 'U238', with a nucleus containing 238 'nucleons' (92 protons and 146 neutrons). The atomic bomb dropped on Hiroshima used a much less common isotope of uranium, U235 – with 92 protons but only 143 neutrons. Naturally occurring uranium contains less than 1% of the U235 isotope.

Confusion about the 'isotopes' discovered by Rutherford and Soddy at the turn of the century persisted for a long time. The mystery was only sorted out after James Chadwick discovered the neutron in 1932. Before Chadwick's discovery increased the number of elementary particles by one, another scientist grasped the significance of Einstein's mass–energy relation for nuclei, and indicated how the energy stored in the nucleus might be released. This was another Cambridge physicist, Francis Aston, who pursued science and sport with equal enthusiasm. Besides skiing and racing motorcycles, he was one of Rutherford's Sunday golf partners in Cambridge. Originally a chemist, he entered the Cavendish Laboratory in 1910 to work with the famous J. J. Thomson on the problem of isotopes of neon. Aston attempted to confirm Thomson's suspicion that there were two isotopes of neon using a method of isotope separation called 'gaseous diffusion'. This technique had been devised by a German scientist, Klaus Clusius. In its sim-

Figure 7.3 Francis William Aston and James Chadwick (right) in the mid-1930s. Aston graduated in chemistry at Birmingham and worked as a chemist for three years in a brewery before turning to physics as a career. From 1909, he worked in Cambridge as J. J. Thomson's assistant. He won the Nobel prize for chemistry in 1922. [University of Cambridge, Cavendish Laboratory.]

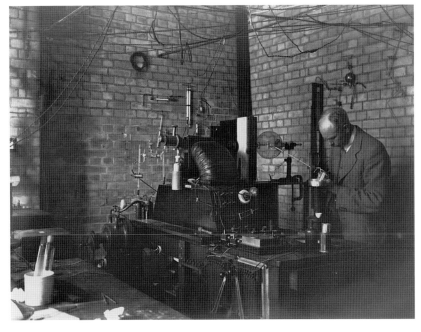

Figure 7.4 The third mass spectrograph and its designer, Francis Aston. The globe contained atoms of the material to be tested either as a gas or as part of the positive terminal. Ions of a particular kind are produced, and the beam is bent first by electric fields and then by magnetic fields produced by coils. A photograph is then produced in which the ions are separated by their mass and charge. [University of Cambridge, Cavendish Laboratory.]

plest form, the necessary apparatus was a long tube standing on end, with a heated wire running down the middle. When the tube was filled with gas, Clusius had shown that the random motion of the gas molecules and gravity could separate different isotopes: the gas at the top became enriched with the lighter isotope and the gas at the bottom contained more of the heavier isotope. The mass difference between isotopes is very tiny on an everyday scale – just a few extra neutrons – so this process has to be repeated many times before any significant separation is achieved. Before Aston had managed to convince Thomson of his results, the war intervened and Aston left to help the war effort. During the war years he continued to think about the problem, and when he returned to the Cavendish, after the war, Aston was convinced that he now knew how to devise a better method of detecting the mass difference. His improvement consisted of projecting a beam of ionized atoms through a combination of electric and magnetic fields. Electric fields deflect charged particles, and magnetic fields cause a beam of charged particles to follow a curved path, and the amount of curvature produced depends on the mass of the isotope. Aston was able to produce separate beams of individual isotopes and correlate different nuclear masses with their arrival positions on a strip of film. The principle of his technique was similar to that of an optical 'spectrograph', which separates different colours of light by their frequency in the same way that a prism produces a rainbow from white light. Thus, by analogy, Aston called his new device a 'mass spectrograph'.

Using this new spectrograph, Aston was able to identify 212 of the 281 naturally occurring isotopes we know today. He was also able to solve Thomson's old problem of neon. He wrote that

Figure 7.5 A modern mass spectrometer in the Physics Department of the University of Wales at Swansea. The laser beam visible in the picture disrupts the target, and ions are accelerated into the tube on the right. The mass of the molecular fragments is distinguished by measuring the time of flight along the tube.

neon consisted, beyond doubt, of isotopes 20 and 22, and that its atomic weight 20.2 was the result of these being present in the ratio of about 9 to 1.

Despite this success, Aston was puzzled. He had expected all the nuclear masses to be simple multiples of the mass of hydrogen. In fact, the so-called 'atomic mass unit' is defined so that the most common isotope of carbon is exactly 12 units. With this definition, hydrogen has a mass of 1.008 units. Why was it not exactly 1 unit? Why was the mass of helium not four times this hydrogen value, 4.032 units, but only 4.002 units? Why was oxygen only 15.994 units and not equal to sixteen hydrogen masses? Aston realized that mysterious 'nuclear forces' – forces much more powerful than the repulsive electric force between the protons in the nucleus – must be holding the nucleus together. Perhaps the hydrogen nuclei had to give up some mass when packed together in a nucleus as the price for nuclear stability? Aston called this difference in masses the 'mass defect': we now call it 'nuclear binding energy'. He realized that mass converted to binding energy was present in all elements. Aston remarked that if four hydrogen atoms could be converted to a helium nucleus a 'prodigious' amount of energy could be generated:

> Thus to change the hydrogen in a glass of water into helium would release enough energy to drive the *Queen Mary* across the Atlantic and back at full speed.

Aston's experiments also revealed further insights into nuclear physics. His graph of nuclear binding energy for the different elements showed that, in principle, there were two ways in which this energy could be liberated. The

first is the hydrogen-to-helium *Queen Mary* example. This process is now
known as 'nuclear fusion'. Fusion combines two light nuclei into a heavier
nucleus. Such processes provide the energy in the stars and our Sun.
Fusion reactions can be reproduced on a smaller scale in the laboratory.
Mark Oliphant, one of Rutherford's last collaborators, tells a story
about Rutherford's intuition. Oliphant was working with Rutherford on
deuterium–deuterium collisions at one of the first particle accelerators. After
a day in the laboratory, they both went home puzzled by the results they had
obtained. In the middle of the night, Oliphant was woken by a telephone call
from Rutherford, who said: 'I know what those particles are. They are helium
of mass 3.' The sleepy Oliphant replied: 'Yes sir, but why do you think they
are helium 3?' Rutherford grunted down the phone: 'Reasons, reasons! I feel
it in my water.' Needless to say, Rutherford was right.

Aston's binding energy curve also revealed a second way in which
nuclear energy might be freed. If a heavy nucleus could be broken into two
lighter, more tightly bound nuclei, it is possible to produce a net gain in binding energy. This process is now called 'nuclear fission' – by analogy with the
splitting of cells in biology. Fission is the basic mechanism for energy production in nuclear reactors and 'atomic' bombs. Fusion reactions are used in
the more powerful 'hydrogen bombs': reactors fuelled by fusion reactions are
still a long way from being commercially viable. As early as 1936, Aston
talked about the ethics of nuclear research:

> There are those about us who say that such research should be stopped by
> law, alleging that man's destructive powers are already large enough. So, no
> doubt, the more elderly and ape-like of our prehistoric ancestors objected to
> the innovation of cooked food and pointed out the grave dangers attending
> the use of the newly discovered agency, fire. Personally I think there is no
> doubt that sub-atomic energy is available all around us, and that one day
> man will release and control its almost infinite power. We cannot prevent
> him from doing so and can only hope he will not use it exclusively in blowing
> up his next door neighbour.

It was fortunate for the world that it was not until the eve of the Second World
War that nuclear fission was discovered – in Berlin.

In a lecture to the Royal Society in June 1920, Rutherford speculated about 'the possible existence of an atom of mass 1 which has zero nucleus
charge'. He pointed out that such a neutral particle, which he called the neutron, 'should enter readily the structure of atoms' and be an exceptionally
useful probe of the atomic nucleus. A young assistant of Rutherford, James
Chadwick, attended this lecture, and was not initially impressed by this idea.
Chadwick was born in the north of England, near Manchester, and it was to
Manchester University that he applied to read mathematics. At the interview
for entrance, Chadwick joined the wrong queue and ended up taking physics
instead. In 1913, after working in Rutherford's group at Manchester,
Chadwick won a scholarship that allowed him to go to Berlin. Rutherford's
collaborator Hans Geiger had returned to Berlin, and Geiger took care to

Figure 7.6 James Chadwick (right) with Peter Kapitza. Chadwick began research with Rutherford in Manchester. He was in Berlin working with Geiger when the First World War broke out, and he was interned in a racehorse stable. After the war, he returned to work with Rutherford in Cambridge. In the mid-1930s, Chadwick wished to build a cyclotron to accelerate particles to higher energies. Rutherford opposed such expense, and Chadwick moved to Liverpool, where he built the first British cyclotron. [AIP Emilio Segrè Visual Archives, Margrethe Bohr Collection.]

introduce the young Chadwick to all the famous Berlin scientists, including Einstein. Unfortunately for Chadwick, his stay in Berlin was involuntarily extended: he was interned in Germany for the duration of the First World War, and he was forced to exchange the laboratory in Berlin for a stable block near Spandau. After the war, Chadwick followed Rutherford to Cambridge and performed many unsuccessful experiments during the 1920s to try to find Rutherford's proposed neutral particle.

The vital clue to discovering the neutral particle was provided by a German physicist, Walther Bothe, with his student Herbert Becker, in 1930. They bombarded beryllium with alpha particles from a radioactive polonium source and found that an intense beam of penetrating neutral radiation was emitted. They interpreted this radiation as high-energy gamma rays, but noted that the energy of these gamma rays was greater than that of the original alpha particles. Conservation of energy requires that this extra energy must come from somewhere, and a nuclear reaction was the likeliest source of energy. In Paris, Marie Curie's daughter, Irene, with her husband Frederic Joliot, decided to use a much stronger polonium source to repeat Bothe's experiment. In their first paper, they confirmed that the mysterious radiation from beryllium had an energy three times that of the incoming alpha particle. Their second paper showed that this neutral radiation was able to eject protons from paraffin wax – a substance rich in hydrogen. Irene Curie and Joliot concluded that the radiation from beryllium was ordinary gamma radiation, i.e. high-energy photons, and that they were seeing protons from the collisions of these gamma rays with the hydrogen atoms in the paraffin wax. Although gamma rays were known to scatter electrons, it did not seem likely that they would be able to scatter protons so strongly, since protons are some 2000 times heavier than electrons. In order to explain Curie and Joliot's interpretation of their experiment, the gamma rays would have to interact with protons more than three million times more strongly than would be anticipated from the electron result. It was not surprising that the physicists in Cambridge were astonished. When told by Chadwick of the Curie and Joliot results, Rutherford abandoned the caution of a lifetime, and said flatly, 'I don't believe it!'

According to the Italian physicist Emilio Segrè, when Ettore Majorana, a young theoretical physicist in Rome, read the French paper he said: 'What fools! They have discovered the neutral proton and do not realize it.' Proving the existence of the neutron beyond all doubt required painstaking work. Chadwick took up the challenge to solve the mystery, and said later, with true British understatement, 'It was a strenuous time.' He continued doing all his administrative jobs at the Cavendish Laboratory but worked on the problem for ten days, averaging about three hours of sleep a night. Chadwick was able to show that the powerful beryllium radiation was able to knock protons not only out of paraffin wax, but also out of helium, lithium, beryllium, boron, carbon, nitrogen, oxygen and argon. A detailed examination of conservation of energy and momentum then showed that the mysterious beryllium radiation could not be gamma rays. Comparison of the energies of the recoiling nuclei showed that the neutral radiation must have

Figure 7.7 This apparatus built by James Chadwick acted as a neutron source. Alpha particles from radioactive material inside the cylinder struck a beryllium target and generated neutrons. The upright tube was fixed to an air pump. [University of Cambridge, Cavendish Laboratory.]

a mass approximately equal to that of the proton! After announcing his results to the Cavendish scientists, Chadwick concluded his presentation with the words: 'Now I want to be chloroformed and put to bed for a fortnight.'

Chadwick's discovery of the neutron in 1932 began a new phase in the exploration of the nucleus. An alpha particle has a positive electric charge and is repelled by the positive charge of the nucleus: a neutral neutron sees no such electromagnetic barrier. The German physicist Hans Bethe, who played a major role in unravelling the nuclear physics of the atomic bomb, once said that he considered everything before 1932 to be '…the prehistory of nuclear physics, and from 1932 on the history of nuclear physics.' The difference was the discovery of the neutron.

The discovery of nuclear fission

During the First World War, Rutherford had carried on with his nuclear research at Manchester almost single-handedly, doing experiments in the spare time left after his war-time work on submarine detection. His colleague, Ernest Marsden, whose alpha particle scattering experiments with Geiger had given Rutherford the clue to discovering the atomic nucleus, had left for New Zealand in 1915. Before he left, Marsden bequeathed Rutherford a puzzle. In his last experiments, Marsden had caused alpha particles from radioactive radon gas to collide with hydrogen atoms in a container. Using a zinc-sulphide scintillation screen as a detector, together with some metal foil absorbers, he was able to measure how far the particles that produced the scintillations recoiled. Since hydrogen atoms recoil about four times as far as the more massive helium atoms, Marsden was able to distinguish between protons and alpha particles. To his great surprise, Marsden found that, after evacuating the container but before filling it with hydrogen, he saw, in the words of Rutherford, 'a number of scintillations like those from hydrogen'. Where did these protons come from? Marsden concluded that the hydrogen nuclei – protons – appeared to come from the radioactive matter. This was a revolutionary idea: up to then only alpha particles, beta particles or gamma rays had been detected from radioactive processes.

Rutherford was suspicious, and he decided to find out the truth in his usual thorough manner. The series of experiments Rutherford performed to track down the source of the mysterious protons show him to be every bit as good a detective as Sherlock Holmes. After methodically eliminating all other explanations, Rutherford was left with a different, but equally startling, solution to Marsden's puzzle. Rutherford had found that filling the container with oxygen or carbon dioxide decreased the number of protons arriving at the scintillator. To his surprise, when he tried filling the container with ordinary air, the number of proton scintillations doubled. Since air is nearly 80% nitrogen, he suspected that nitrogen must be the culprit. To confirm this, he repeated the experiment using pure nitrogen, and his suspicions were confirmed when he saw the same doubling effect. These results suggested that the hydrogen came from the nitrogen, not from the radioactive source. After carefully checking that this was so, Rutherford had the final piece of the puzzle. He wrote up his results in a series of papers entitled 'Collisions of alpha particles with light atoms'. The last paper in the series was subtitled 'An anomalous effect in nitrogen', and it contained the statement:

> We must conclude that the nitrogen atom is disintegrated under the intense forces developed in a close collision with a swift alpha particle, and that the hydrogen atom liberated formed a constituent part of the nitrogen nucleus.

Figure 7.8 Ernest Orlando Lawrence (1901–1958), inventor of the cyclotron. Lawrence is holding the first particle accelerator in his hand: later examples were gigantic by comparison. The cyclotron was the first accelerator worthy of the name 'atom smasher', a term much loved by the popular press. [Ernest Orlando Lawrence Berkeley National Laboratory.]

The newspapers were more plain-speaking: headlines announced that Rutherford had 'split the atom'. Rutherford had observed the first artificial nuclear reaction or transmutation. An alpha particle, atomic weight 4, collides with a nitrogen nucleus, atomic weight 14, and a hydrogen nucleus, or proton as Rutherford later called it, with atomic weight 1, is ejected, leaving an isotope of oxygen, atomic weight 17, as the recoiling particle. The *nuclear* reaction is written as

$$He^4 + N^{14} \rightarrow O^{17} + H^1$$

Because the alpha particles are positively charged, they are repelled by the positively charged nitrogen nuclei and only very few of the alpha particles are able to penetrate the nuclei to cause the nuclear reaction. Rutherford concluded his 1919 paper with the words:

> …if alpha particles – or similar projectiles – of still greater energy were available for experiment, we might expect to break down the nuclear structure of many of the lighter atoms.

Figure 7.9 The '11 inch' cyclotron was completed by Lawrence and Livingston in January 1932. The two coils – the drum-like structures – provide a magnetic field, which keeps the protons moving in a circular path between the magnets. The protons are accelerated by a radio frequency electric field applied across two hollow D-shaped vacuum chambers in which the protons circulate. Protons of an energy comparable with that of radioactive particles could be produced by this device. Bigger machines were built in rapid succession. There was a problem when the particles reached relativistic speeds and the machines failed to work. New machines called synchrocyclotrons, such as the Bevatron (Figure 6.18), which took account of the relativistic mass increase at high speeds, were needed. [Ernest Orlando Lawrence Berkeley National Laboratory.]

This remark foreshadowed the future progress in nuclear physics through the construction of ever more powerful particle accelerators. With Rutherford's encouragement, John Cockcroft and Ernest Walton, at the Cavendish Laboratory in Cambridge, were the first to produce a nuclear reaction using artificially accelerated protons. Rutherford disapproved of expensive experiments, and, in the true Cambridge 'string and sealing wax' tradition, Cockcroft and Walton's machine was cobbled together out of car batteries, parts from petrol pumps and putty. An American physicist from South Dakota, Ernest Lawrence, who thought on a much grander scale, pioneered the first steps towards our present-day particle accelerators. Together with the theorist J. Robert Oppenheimer, Lawrence established Berkeley in California as one of the great world centres of physics. His 'cyclotron' machine, invented in the early 1930s with help from his student Stanley Livingstone (the exact amount of help remains a matter of controversy), used a combination of electric and magnetic fields to accelerate charged particles. Nowadays, the very much more powerful accelerators in operation at CERN in Geneva and at FermiLab near Chicago accelerate particles to such high speeds that Einstein's predicted relativistic mass increase must be taken into account. The daily operation of these accelerators is a witness to the success of Einstein's modification of the classical Newtonian physics.

Figure 7.10 The Italian physicist Enrico Fermi (1901–1954) was one of the greatest scientists of modern times and was one of the few who made significant contributions to both experimental and theoretical physics. Fermi's wife was Jewish, and he used the opportunity of the 1938 Nobel prize to escape from Mussolini's Italy to America via Sweden. [Brown Brothers.]

Despite the gradual slide towards war in Europe and the grave economic troubles of many nations, the 1930s was an exciting time for nuclear physics. After narrowly failing to discover first the neutron and then the positron, Frederic Joliot and Irene Curie finally made the discovery that won them their Nobel Prize – in 1934 they discovered 'artificial' radioactivity. Using alpha particles to bombard an aluminium target, they had found a new radioactive isotope of phosphorus. This discovery, combined with the use of the neutron as a new neutral probe unaffected by the charge barrier of the nucleus, opened the way for nuclear fission and, subsequently the atomic bomb. Marie Curie died in 1934, but she lived long enough to see her daughter's triumph. Joliot wrote:

> Marie Curie saw our research work and I will never forget the expression of intense joy which came over her when Irene and I showed her the first artificially radioactive element in a little glass tube. I can still see her taking in her fingers (which were already burnt with radium) this little tube containing the radioactive compound – as yet one in which the activity was very weak. To verify what we had told her she held it near a Geiger–Muller counter and she could hear the rate meter giving off a great many 'clicks'. This was doubtless the last great satisfaction of her life.

Before she died, Marie Curie managed to insert a last-minute revision describing this new discovery into the new edition of her book on radioactivity.

It was the Italian physicist, Enrico Fermi, who next took up the challenge. Fermi realized that neutrons could initiate nuclear reactions much more easily than charged particles because they are not repelled by the positive electrical charge of the nucleus. He had assembled a talented team of young physicists in Rome, and they started a systematic exploration of the Periodic Table. Their first paper reported a new radioactive element produced by bombarding aluminium with neutrons. This was rapidly followed by a second paper reporting the discovery of artificially induced radioactivity in sodium, magnesium, silicon, phosphorus, chlorine, titanium, vanadium, chromium, iron, copper, zinc, arsenic, selenium, bromine, silver, antimony, tellurium, iodine, barium and lanthanum, in order of increasing atomic number. In the spring of 1934, they reached the heaviest known element, uranium, and found signals for new radioactive elements. The different radioactive isotopes produced were distinguishable by their different 'half-lives' – the half-life of a radioactive element measures how long it takes for the radioactivity to 'decay' to half its original value. After neutron bombardment, it was often very tricky to sort out what was actually left behind. U238 (atomic number 92) is naturally radioactive and decays through a series of fourteen steps down the Periodic Table to finish as lead (Pb, atomic number 82).

Figure 7.11 The Periodic Table.

Fermi realized that adding a neutron to U238 could produce the new isotope

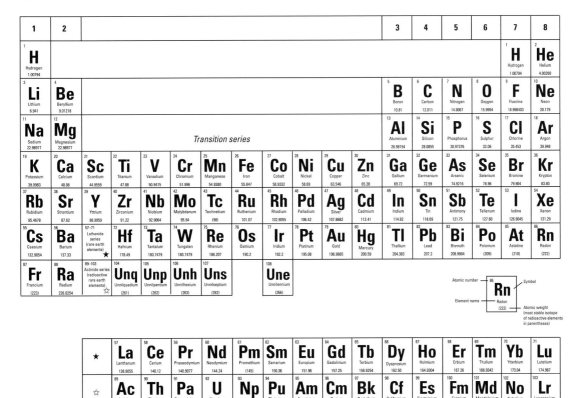

U239. This isotope could then beta decay – a process equivalent to a neutron ejecting an electron and turning into a proton. This would create a new man-made element containing 93 protons. By June, they had done enough checks for Fermi to suggest in print that 'the atomic number of the element may be greater than 92'. In fact, they were wrong.

By 1938, racism and anti-semitism had spread to Italy. Fermi's wife, Laura, was Jewish, and so the couple decided to emigrate as soon as possible. Niels Bohr sympathized with their plight, and deliberately broke with tradition by hinting to Fermi that he would probably win that year's Nobel Prize. This gave the Fermis an opportunity for escape: since they could not take all their belongings with them for fear of alerting the authorities, the prize money would make it possible for them to start afresh in America. Fermi had only just arrived in New York when he heard news that forced him to add an embarassed footnote to the Nobel Prize lecture he had given only weeks before. Otto Hahn and Fritz Strassmann, working at the Kaiser Wilhelm Institute in Berlin, had been attempting to disentangle the mixture of products from neutron bombardment of uranium by chemical means. To their great confusion and surprise, they had identified barium – with atomic number 56 – amongst the decay products. They announced their reluctant conclusions in a paper:

> As a consequence of these investigations we must change the names of the substances mentioned in our previous disintegration schemes, and call what we previously called radium, actinium, and thorium, by the names barium, lanthanum and cerium. As nuclear chemists who are close to the physicists, we are reluctant to take this step that contradicts all previous experiences of nuclear physics.

Because of the worsening political situation in Germany, Hahn's long-time collaborator, Lise Meitner, had had to leave her post as head of physics at the Kaiser Wilhelm Institute and flee to Holland. Meitner and Hahn had spent many years puzzling over the physics and chemistry of radioactive decay chains, so it was natural that Hahn should first write to her for advice:

> Perhaps you can suggest some fantastic explanation. We understand that it really can't break up into barium... .

Figure 7.12 Lise Meitner (1878–1968) studied physics with both Boltzmann and Planck. She became interested in radioactivity and worked in Berlin with Otto Hahn for thirty years. In the late 1930s it became too dangerous for a 'Jewish Protestant' to stay in Germany. Although she had made decisive contributions that established the possibility of a nuclear chain reaction, she declined to work on the atomic bomb. [Churchill Archives Centre, Churchill College, Cambridge and Meitner-Graf Studio Portraits.]

It was Christmas, 1938, and Meitner was in Sweden on holiday with her nephew, Otto Frisch. Before he received any reply, Hahn had confirmed the presence of lanthanum – a neighbour to barium in the Periodic Table with atomic number 57. He wrote Meitner a second letter explaining that he felt he had to publish quickly: 'We cannot hush up the results even though they may be absurd in physical terms.' He hoped that Meitner would provide an acceptable explanation, which she did. Meitner and Frisch realized that, rather than creating new 'trans-uranic' elements, they were observing the uranium nucleus breaking up into two big pieces – 'nuclear fission'. Frisch went back to Copenhagen, where he quickly performed an experiment showing conclusive evidence for this. As soon as he had returned to Copenhagen, Frisch told Bohr about their 'fission' idea:

> I had hardly begun to tell him, when he struck his forehead with his hand and exclaimed, 'Oh what idiots we have all been! Oh but this is wonderful! This is just as it must be!'

So it was Bohr, who was leaving for the USA in a few days time, who brought news of the discovery of fission to the new world.

There are two interesting footnotes to this story that demonstrate how a combination of luck, energy and intelligence is required in physics. In 1935, a German chemist named Ida Noddack wrote a paper entitled 'On element 93'. In it, she specifically pointed out that Fermi's group in Rome had not proved that the uranium nucleus did not break up into two large fragments. This suggestion was so out of line with current theoretical thinking that it was uniformly ignored. The Rome group had enshrined this thinking into one of their key assumptions:

> It is reasonable to assume that the atomic number of the active element should be close to the atomic number ... of the bombarded element.

Figure 7.13 Otto Robert Frisch (1904–1979). Frisch, with his aunt Lise Meitner, was the first to understand nuclear fission. With Rudolf Peierls he wrote the famous Frisch–Peierls memorandum, which showed for the first time that an atomic bomb was feasible. Despite having written this memorandum and then passing it to the British government, there was initially an absurd attempt to exclude both Frisch and Peierls from further discussions of the bomb on the grounds that they were not British. [Courtesy Mrs U. Frisch.]

Unfortunately, Noddack did not follow up her suggestion by performing the relatively simple experiments to confirm her idea. Another cautionary tale of the undue influence of theoretical prejudice comes from the Cavendish Laboratory in Cambridge. Scientists repeating Fermi's experiments with uranium saw strong pulses caused by the fission fragments, but thought that they were due to faulty detectors! We should, however, probably be grateful that the discovery of fission was delayed until the onset of the Second World War. Had it been discovered a few years earlier, the customary free exchange of scientific results would have meant that the nuclear physics of fission would have been well explored by the time the war began in earnest. Everyone would have realized what was needed to make a bomb, and there seems little doubt that Hitler would have had no scruples about using such a 'wonder' weapon. H. G. Wells' terrifying scenario about a Europe devastated by atomic bombs might then have become reality.

Tube Alloys and the Manhattan Engineering District

The idea of using nuclear reactions to produce energy – either peacefully in a reactor or explosively in a bomb – had been considered long before the discovery of fission. Recall that, in 1934 Leo Szilard, living in a London hotel, in exile from his native Hungary, had even patented the idea. He realized that neutrons might be better than charged particles at starting nuclear reactions, and that there was the possibility of a 'chain reaction'. If an element could be found that released two or more neutrons for each one captured, the new neutrons could themselves react and produce more neutrons. This is the basic feature of a 'chain reaction' – an avalanche of reactions originated by a single neutron. Szilard's 1934 patent described this as follows:

> A chain reaction in which particles which carry no positive charge and the mass of which is approximately equal to the proton mass... form the links of a chain.

He also foresaw that there would need to be a 'critical mass' (as it was later called) of the reacting element in order for a chain reaction to be maintained because neutrons can be lost in various ways. One way of losing neutrons is by escape from the surface, so Szilard proposed reducing the critical mass required by surrounding the chain-reacting element by 'some cheap heavy material, for instance lead', which would reflect the escaping neutrons back into the reacting core. This idea later found its way into the nuclear bomb programme under the name of 'tamper', by analogy with the mud 'tamped' into the drill holes when conventional explosives are used. Szilard was under no illusions as to what could happen if he managed to assemble more than a critical mass of material: 'If the thickness is larger than the critical value... I can produce an explosion.'

Although Szilard, and others, had correctly recognized the possibility of self-sustaining nuclear reactions, there was much nuclear physics to be understood before such an idea could become a reality. Precise knowledge of how effective neutrons of different energies were in causing fission to take place, and the number and timing of the extra neutrons produced, turned out to be essential. Emilio Segrè, the Italian Nobel Prize winner who worked on the bomb project during the Second World War, summed up the situation memorably:

> In an enterprise such as the building of the atomic bomb the difference between ideas, hopes, suggestions and theoretical calculations, and solid numbers based on measurement, is paramount. All the committees, the politicking and the plans would have come to naught if a few unpredictable nuclear cross-sections had been different from what they are by a factor of two.

The term 'cross-section' is used by nuclear physicists to describe the probability that a nuclear reaction will take place: it is a measure of the effective area of the target nucleus presented to the incoming particles for a particular reaction. The larger the cross-section, the higher the probability of the reaction. One of the key results about neutron reactions had been discovered by Fermi and the Rome group well before the discovery of fission. The manner of this discovery also says something about the mysterious 'sixth sense' possessed by all great physicists.

Hans Bethe, J. Robert Oppenheimer's chief theoretician in the atomic bomb programme at Los Alamos, once said that one of the key results of neutron physics 'might never have been discovered if Italy were not rich in marble'. Fermi's group had found that they obtained different results for their neutron scattering experiments carried out on wooden and marble tables – not a difference likely to have been noticed on the wooden benches of the Cavendish Laboratory. Fermi decided to repeat the experiment with a piece of lead inserted between the neutron source and the target. After machining a piece of lead to exactly the right shape, Fermi felt a strange reluctance actually to put the lead in place. Finally, he said to himself, for no conscious reason: 'No, I do not want this piece of lead here; what I want is a piece of paraffin.' The result was a dramatic increase in the amount of radioactivity produced – so dramatic, that Segrè, one of Fermi's team, at first thought that one of their radioactivity counters must have gone wrong. In true Italian fashion, the discovery did not prevent Fermi going home for lunch: he returned later that day with the explanation. Because they had to get past the electrical barrier to reach the nucleus, fast alpha particles and protons were more effective than slow ones in starting a nuclear reaction. Fermi realized that this was not true for the uncharged neutrons. Slowing down the neutrons before they reached the target nucleus gave the neutrons more time near the nucleus and increased the likelihood of a nuclear reaction. What was happening in their puzzling experiments was that the neutrons were being slowed down more by collisions with hydrogen atoms in the wood and the paraffin than by collisions with the calcium and oxygen nuclei in the marble.

In the case of nuclear fission, the material chosen to slow down fast neutrons produced at each stage of the chain reaction is called the 'moderator'. Heavy nuclei are much less effective at slowing down neutrons than light nuclei. Since neutrons weigh almost the same as protons, water would seem a good first choice as a moderator. Unfortunately, besides slowing down the neutrons, water also absorbs some neutrons by a nuclear reaction that produces 'heavy water'. Heavy water still has the chemical formula H_2O but the hydrogen is in the form of 'heavy hydrogen', or deuterium – an isotope of hydrogen with a neutron and a proton as its nucleus. In Nature, heavy water is a very rare component of ordinary water. If the percentage of heavy hydrogen in the water could be increased, heavy water would be a very efficient modera-

Figure 7.14 The German-American physicist Hans Bethe (right) with Isidor Rabi. Bethe proposed the first detailed scheme for the series of nuclear reactions taking place inside stars. In his theory, four hydrogen nuclei are converted to a helium nucleus with the aid of a series of catalytic reactions involving carbon 12. This carbon cycle is believed to be the dominant process inside more massive stars. [Photograph by S. A. Goudsmit. Courtesy AIP Emilio Segrè Visual Archives.]

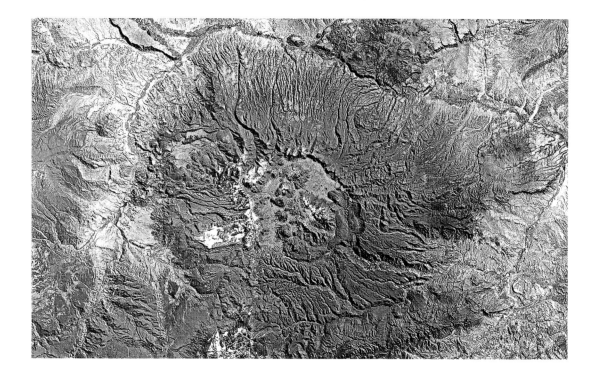

Figure 7.15 Landsat photograph of the remote area of New Mexico which General Leslie Groves and J. Robert Oppenheimer decided to make the centre of the US atomic bomb programme. Los Alamos Laboratory is located between the Rio Grande at the top and the snow-capped mountains of the Valles Caldera, an extinct volcano some 15 miles in diameter.

tor of fast neutrons produced in a uranium reactor. An alternative to heavy water would be a light element that has a small cross-section for absorbing fast neutrons. Fermi and Szilard, both European exiles working in the United States, chose to use carbon as the moderator for the world's first self-sustaining nuclear reactor. The German physicists – Heisenberg among them – chose heavy water. One of the most successful sabotage operations of the Second World War was the destruction of the German stocks of heavy water produced by the Vemork plant in occupied Norway. The heavy water was being transported to Germany under guard, and the Norwegian Resistance, with help from British Intelligence, decided that the only practical possibility was to sink the ferry that would carry the water from the plant to the rail-head. The resistance team planted a home-made time-bomb on the ferry, and the sabotage operation went, quite literally, like clockwork. At 10.45 a.m. on Sunday, February 20th, 1944, the bomb exploded and the ferry was sunk with the loss of twenty-six lives. Kurt Diebner of the German Army Ordnance said after the war:

> It was the elimination of German heavy water production in Norway that was the main factor in our failure to achieve a self-sustaining atomic reactor before the war ended.

Why was it important to build a nuclear reactor if the aim was to build a bomb? The answer to this question lies in the very different behaviour of the two isotopes U238 and U235. U235 fissions very easily with slow neutrons, but is present in natural uranium at less than the 1% level: U238 is, for all practical purposes, not fissionable. What then happens to the U238 when it reacts with a fast neutron? There is a high probability for capture of the neutron to form the isotope U239. This then decays, ejecting an electron to form a new, genuine, 'trans-uranic' element. In 1940, the American physicist, Louis Turner from Princeton, speculated not only that 'element 93' could be produced in this way, but also that it too would probably beta decay, ejecting an electron to become 'element 94'. Turner thought that this element would be likely to fission easily, just like U235. In the USA, Szilard had been urging physicists not to publish important results on fission for fear of helping Hitler manufacture an atomic bomb. After submitting his paper to the journal *Physical Review*, Turner had second thoughts and asked Szilard for advice about publication. Although Turner thought his paper 'was wild enough speculation so that it could do no possible harm', Szilard thought otherwise. In the event, Turner's paper was not published until 1946. Szilard was right: he had realized that, although it was very difficult to separate U235 from U238, since the isotopes are chemically identical, element 94 would be a new easy-to-fission element that could be chemically separated with relative ease. Carl von Weizsacker, one of the key figures in the German 'Uranium Club' that formed the core of the German bomb programme, had had the same idea. In a paper dated July 17th, 1940, Weizsacker summarized his conclu-

Figure 7.16 The K-25 gaseous diffusion plant built at Oak Ridge, Tennessee, to separate fissionable U235 from the dominant U238. The structure is half a mile long. [Martin Marietta Systems Inc.]

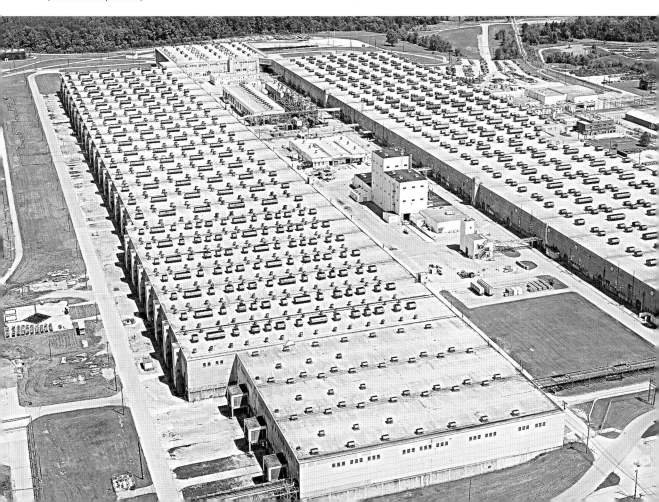

sions: '…neutron bombardment of uranium in a chain-reacting nuclear reactor would produce a new element that was easily separable and which could be used in a bomb.'

Niels Bohr, who had been first to recognize the crucial role played by U235 in fission, believed an atomic bomb to be impossible because of the difficulty of separating U235 from U238. The existence of element 94 – referred to 'in code' throughout the war as '49' at Los Alamos – completely changed the picture. The last paper to be published on fission research until the end of the war appeared in June 1940, and was entitled 'Radioactive element 93'. Edwin McMillan and Philip Abelson, working at Berkeley in California, had painstakingly established that U239, produced by neutron capture, decayed to a new chemical element that itself also then decayed. McMillan had also thought of a name for their new element – neptunium – since Neptune is the next planet out from Earth after Uranus. Although McMillan suspected he had also produced element 94, he was unable to prove this conclusively. McMillan's work was continued by a young chemist, Glenn Seaborg, working with a team that included Emilio Segrè from Fermi's group in Rome. In February 1941, Seaborg and his team identified element 94, and a month later confirmed that it 'undergoes fission with slow neutrons'. Although element 94 was not officially named until 1942, Seaborg already had a name for it. Following the sequence from Uranus to Neptune, he proposed to call it plutonium – after Pluto, the planet, but also after the Greek god of the dead.

In spite of Szilard's efforts, the US bomb programme had stagnated in bureaucracy and committees. Perhaps because the threat of Nazi domination seemed more real to them, it was two physicist refugees in England who made the crucial next step. At Birmingham University in March 1940, Otto Frisch asked Rudolf Peierls the question: 'Suppose someone gave you a quantity of pure 235 isotope of uranium – what would happen?' Peierls had developed a formula to calculate critical masses, so they put in the numbers for U235. They were amazed at how small a mass was needed:

> We estimated the critical size to be about a pound, whereas speculations concerned with natural uranium had tended to come out with tons.

In fact, their number turned out to be a little on the low side because they had had to make a plausible guess about the fission cross-section of U235 – but they were right that only pounds and not tons were needed. Could such a chain reaction lead to an explosion or would the uranium merely 'fizzle'? The chain reaction has to proceed very quickly, otherwise the pressure caused by the fissioning atoms will push the uranium atoms too far apart for the chain reaction to continue. Peierls made a rough estimate, 'on the back of the proverbial envelope', assuming that fast neutrons also caused U235 to fission. His calculation showed that some eighty links of the chain would be generated before the pressure blew the uranium apart. This meant that a pound or so of uranium would release the equivalent energy of thousands of tons

Figure 7.17 Richard Feynman at Los Alamos, a brilliant young group leader and celebrated safe-cracker: 'I opened the safes which contained behind them the entire secret of the atomic bomb...'. [Courtesy Michelle Feynman.]

of ordinary explosive – the kiloton bomb had arrived. They were in awe of their own results. Separation of tons of U235 was not a practical proposition: separation of a few pounds was. Frisch had been working on the Clusius method of isotope separation, and he estimated that, with a separation plant with a chain of 100 000 such tubes, 'one might produce a pound of reasonably pure uranium-235 in a modest time, measured in weeks'. Frisch and Peierls said to themselves: 'Even if this plant costs as much as a battleship, it would be worth having.'

Despite some bureaucratic absurdities – at first, Frisch and Peierls were thanked for their efforts but told that as actual or former 'enemy aliens' they could not be told anything more about it – the British were much quicker than the Americans to put in place a serious bomb programme. After asking Mark Oliphant, head of physics in Birmingham, for advice, Frisch and Peierls wrote up their discoveries in a two-part report, known later as the 'Frisch–Peierls memorandum'. Since they did not dare give it to a secretary, Peierls had to type the report himself. The memorandum is an amazingly far-sighted document, with the second, less technical part written in such direct and simple language that not even the politicians or the military could fail to grasp its message. After pointing out that the explosion would be large enough to destroy 'the centre of a big city', the authors went on to explain the

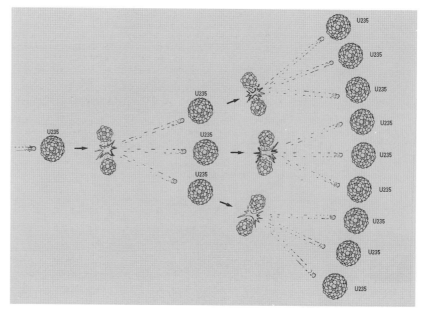

Figure 7.18 A schematic diagram of a chain reaction. The U235 nucleus on the left absorbs a neutron and fissions, producing three more neutrons. These neutrons then cause further fission reactions in other U235 nuclei.

Figure 7.19 Genia and Rudolf Peierls. With Otto Frisch, Peierls wrote a famous memorandum that clearly showed the feasibility of a fast-fission uranium bomb. Although it was their calculations that convinced the British government to initiate its atomic bomb programme under the leadership of the recently formed MAUD committee, for a short time after their memorandum, Frisch and Peierls were excluded from discussion of their own 'secrets' on the grounds that they were enemy aliens! Peierls co-opted another German refugee, Klaus Fuchs, to work with him in Birmingham on the problem of separation of U235 from U238. Fuchs also worked with Peierls at Los Alamos, and Peierls felt a great sense of personal betrayal when Fuchs was arrested in 1950 for spying for the Russians. [Courtesy Jo Hookway.]

implications of the radioactive contamination such a bomb would cause.

Peierls and Frisch ended their report with a bleak warning about a possible German bomb:

> Since the separation of the necessary amount of uranium is, in the most favourable circumstances, a matter of several months, it would obviously be too late to start production when such a bomb is known to be in the hands of Germany, and the matter seems, therefore, very urgent.

Frisch and Peierls were right to be worried. After the war was over, it was discovered that in December 1939, Heisenberg had written a report for the German War Office in which he concluded that 'enriched uranium' with more than the normal percentage of U235 was 'the only method of producing explosives several orders of magnitude more powerful than the strongest explosives yet known'. Fortunately for the allies, the German War Office did not follow up Heisenberg's report. After the defeat of Germany, Heisenberg and other German nuclear physicists were interned in England at Farm Hall near Cambridge. All their conversations were secretly recorded including their reaction to the BBC news on the radio that the Americans had dropped an 'atomic bomb' on Hiroshima. Their shock and disbelief displayed in the transcripts make clear that they had no idea that the critical mass of U235 could be so small. A hotly debated question amongst physicists and historians of science is whether Heisenberg deliberately misled the German bomb programme. Peierls, at least, refused to countenance such a suggestion, saying grimly that 'those who choose to sup with the devil had better make sure they have a long enough spoon!' Carl von Weisacker, one of the leaders of the

Figure 7.20 Emil Klaus Fuchs was one of the most celebrated 'atom spies'; the information he passed to the Russians shortened the development time of their nuclear and hydrogen bombs by a significant amount. Fuchs was born in 1911 near Darmstadt in Germany and, unlike the vast majority of German refugees, was not Jewish. He was a member of the Communist party at Kiel University, where he was beaten up by Nazi student Brownshirts and thrown into the river. After the Reichstag fire, the Nazis blamed the Communists, and Fuchs fled to Switzerland. In 1933 he arrived in England and became Nevill Mott's first PhD student in Bristol. He was recruited by Peierls for the British bomb programme, and followed Peierls to Los Alamos. From 1942 to 1950 Fuchs passed on very detailed information, not only about the uranium and plutonium bombs, but also about the 'Super' – Teller's hydrogen bomb project. With John von Neumann, Fuchs was the author of the top secret 'Disclosure of Invention' or patent that summarized all the significant progress towards hydrogen fusion bombs. After the war, Fuchs returned to England to become head of the theoretical physics division at the newly formed Harwell Atomic Energy Laboratory. An American controller arranged that Fuchs should meet his new Russian contact on the first Saturday of every month at the entrance to Mornington Crescent underground station: he would be carrying five books in one hand tied with string and two books in the other. In the USA, suspicion about a security leak had been narrowed down to Peierls or Fuchs: an intercepted message between London and Moscow pointed the finger at Fuchs, who later confessed.

German bomb effort, was invited to lecture in Oxford University long after the war. Peierls, Professor of Theoretical Physics at Oxford, was asked if he would like to attend the reception for von Weisacker. He replied that if he had to speak to von Weisacker he would be polite, but would very much prefer that he was not asked to the reception.

Frisch and Peierls gave their report to Oliphant, who 'promised to get it to the right person'. He evidently succeeded: by June 1940, a group called the MAUD Committee had been set up. Margaret Gowing, Professor of the History of Science in Oxford, has said that MAUD was 'one of the most successful committees this or any other country has ever seen'. The MAUD Committee was thought by some of the people who knew of its existence to be an acronym standing for 'Military Applications of Uranium Disintegration'. In reality, the committee's name had a different origin. Lise Meitner had been in Copenhagen when the Germans invaded, but she had been allowed to return to Stockholm. Before she left, Niels Bohr asked her to send a telegram to a friend in England to say he was all right. She did so, and her telegram ended: 'PLEASE INFORM COCKCROFT AND MAUD RAY KENT'. When Cockcroft received the message he decided the mysterious 'MAUD RAY KENT' must be an anagram for RADYUM TAKEN. This confirmed his worst fears that the Germans were getting hold of all the radium they could find. The truth is more prosaic. Meitner's message had been garbled: Maud Ray was an old governess of the Bohr family, and her full address, in Kent, had been lost in transmission. In the summer of 1941, the MAUD Committee wrote its final report, concluding that an atomic bomb was feasible and outlining what was needed to produce one. The MAUD Report directly led to the setting up of a British atomic bomb programme. This programme was given the code name 'Tube Alloys', and this pseudonym for the atomic bomb programme was used in negotiations between Churchill and Roosevelt about UK–USA collaboration.

It was the MAUD Report that finally triggered the Americans into serious action. The 'Manhattan Engineering District', the code name for the US atomic bomb programme (more usually known as the Manhattan Project), was set up in 1942, and its results are surely known to everyone. 'Little Boy' was the code name for the U235 'gun' bomb dropped on Hiroshima; 'Fat Man' was the name for the plutonium 'implosion' bomb

Figure 7.21 The first man-made nuclear explosion: a sequence of photographs of the first 2 seconds of the Trinity Test. The test took place in southern New Mexico, around 60 miles northwest of Alamogordo, in a region appropriately called Journada del Muerto – the Journey of Death. The explosion was subsequently measured to have been equivalent to 18.6 kilotons of TNT. [Los Alamos Historical Museum Photo Archives.]

Figure 7.22 Robert Oppenheimer and General Leslie Groves visiting the Trinity site after the explosion. The test tower, hundreds of feet of steel girders, winch and shack had all been vaporized apart from a few twisted remains. Instead of asphalt, there was now only green, glassy, fused desert sand. [Los Alamos Historical Museum Photo Archives.]

dropped on Nagasaki (see Figure 7.26). The story of the Manhattan Project and of the Los Alamos Laboratory led by J. Robert Oppenheimer and General Leslie Groves still retains its fascination fifty years on. Incomparably the best account of these times is given by Richard Rhodes in his book *The Making of the Atomic Bomb*. The story reads like an epic novel, with a wealth of sub-plots, ranging from Fermi and Szilard's struggles to construct the first nuclear reactor, to Edward Teller's obsession with the possibility of the 'Super', the hydrogen fusion bomb. There are also many intriguing personal stories about Oppenheimer, Teller, Bethe, Feynman, Peierls and the spy Klaus Fuchs, to name only a few among many. Relationships developed at Los Alamos laid the basis for the rapid expansion of US science after the war.

It is easy now to dismiss as paranoia the US preoccupation with communist spies during the post-war years. In fact, as Richard Rhodes reveals in *Dark Sun*, the sequel to his book on the atomic bomb, extensive Soviet espionage by Klaus Fuchs and others had given the Russians a very detailed account of the work of the Manhattan Project. Stalin and Beria were suspicious that the information was intended to lure them into a costly waste of precious Russian resources. Not until after Hiroshima did Stalin authorize a full-scale project with Beria as its head. Joe-1, the first Russian atomic bomb, exploded in 1949, was a direct copy of Fat Man.

Einstein, almost alone of emigré physicists in the USA, played no active part in the development of the atomic bomb – but it was his fundamental insight of the equivalence of mass and energy that had made it all possible. Einstein spent the better part of thirty years searching for a 'unified field theory' that would encompass both electromagnetic and gravitational forces. It is curious that he should have so completely ignored the developments in nuclear physics outlined in this chapter and the discoveries about the nature of the strong and weak forces.

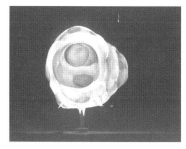

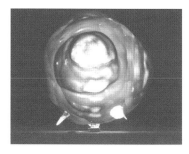

Figure 7.23 A surreal sequence of 'Rapatronics' photographs of the early fireball of the 'TUMBLER/SNAPPER HOW' nuclear test of June 5th, 1952. The Rapatronics pictures reveal details of the explosion at time intervals too short to be seen by the human eye. To reduce fallout, the 14 kiloton device was located on top of a 300 feet high tower: the downward-extending 'spikes' are the shock wave running down the steel guy ropes. [Los Alamos National Laboratory.]

Figure 7.24 A diagram showing an outline of the probable Teller–Ulam configuration for the hydrogen bomb. A Los Alamos publication on the laboratory's fortieth anniversary described the general mechanism as follows: 'The first megaton-yield explosives (hydrogen bombs) were based on the application of X-rays produced by a primary nuclear device to compress and ignite a physically distinct secondary nuclear assembly.' A fusion weapon such as this can be made arbitrarily large by adding more fuel.

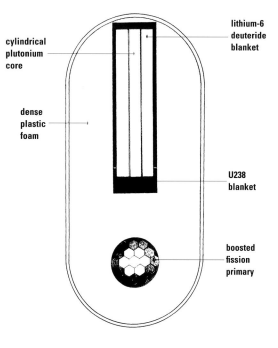

cylindrical
plutonium
core

lithium-6
deuteride
blanket

dense
plastic
foam

U238
blanket

boosted
fission
primary

Figure 7.26 The ruins of Nagasaki. [Los Alamos Historical Museum Photo Archives.]

Figure 7.25 A set of five time-sequence high-speed optical photographs showing a hollow pipe being compressed by a surrounding explosive charge. The bright areas correspond to locations where the interacting shock waves have increased the temperature of the detonation gases. In an implosion bomb, a spherical core is compressed by the shock waves from carefully shaped explosive charges quickly enough so that an explosive chain reaction can be initiated. [Los Alamos National Laboratory.]

8   Down to earth

In 1907... I realised that all natural phenomena could be discussed in terms of special relativity except for the law of gravitation. I felt a deep desire to understand the reason behind this... .

Albert Einstein, Kyoto Lecture, 1922

The weight of light

What do colleagues say about giving up the principle of the constancy of the velocity of light? Wien tries to help himself by questioning the gravitational [action of] energy. That, however, is untenable ostrich politics.
Albert Einstein, letter to L. Hopf, 1912

In 1913, Max Planck visited Einstein in Zurich with the aim of persuading him to move to Berlin. In conversation, Einstein remarked to Planck that he was working on a new theory of gravity. Planck's response was forthright, but concerned:

> As an older friend I must advise you against it, for in the first place you will not succeed; and even if you do suceed, no one will believe you.

Planck was only partly right. Einstein succeeded and his theory of 'general relativity' was believed, but, for the most part, his theory had little relevance to mainstream physics. It was not until after Einstein's death in 1955 that the new technological advances of the 1960s rekindled interest in general relativity. Whereas most advances in understanding Nature could have been made by several scientists working at the time instead of the actual discoverer, this is probably not true of general relativity. Without Einstein's inspiration and persistence, it is likely that theoretical physicists would have taken

Figure 8.1 Planck and Einstein in Berlin in July, 1929. Four years later, the Nazis came to power and Einstein left Europe to spend the rest of his life in America. [AIP Emilio Segrè Visual Archives.]

many more years to arrive at our present understanding of space, time and gravity.

Special relativity emerged from attempts to understand the nature of light and to resolve well-known problems with the theory of electricity and magnetism. Why should gravity be more difficult to reconcile with special relativity? Let us look at the similarities and differences of the forces of electricity and gravity. Gravity is the force we are most familiar with: it is directly experienced as weight. Electricity is less well-known, despite the fact that it is much more powerful. Only rarely do we see this power exposed, as, for example, in the lightning that accompanies a violent thunderstorm. More often, the power of electrical forces is concealed in matter that has no overall electric charge – matter that is electrically neutral. As we have seen in the previous chapters, a neutral atom contains a tiny, positively charged nucleus whose charge is exactly cancelled out by its accompanying swirl of negatively charged electrons. With electricity, like charges repel and unlike charges attract: with gravity, every piece of matter attracts every other piece of matter. At large distances from a massive object, the gravitational force becomes very small, but never disappears entirely. With electrical forces, the situation is different. Michael Faraday showed how electrical forces could be shielded by sitting inside a metal cage. The 'Faraday cage' was charged to such a high voltage that lightning-like flashes sparked from the metal bars. Inside the cage, Faraday experienced no electrical forces, in spite of the dramatic effects occurring outside. So far as we know, it is impossible to build a similar gravitational-force shield.

We experience the force of gravity in our daily activities: the extra energy we need to climb stairs; the ability to play ball games without the ball spinning off into outer space – these are in some part due to the gravitational force exerted on us and things around us by the Earth. Yet the electrical forces hidden inside electrically neutral matter are much stronger than

Figure 8.2 Benjamin Franklin (1706–1790) indulging in a spectacularly dangerous experiment to demonstrate the electrical nature of lightning – several other scientists were killed attempting similar experiments. Franklin was a successful printer and journalist, and only became interested in electricity when he was about forty years old. He is credited with the idea that electric charge can never be created or destroyed, but that it comes in two varieties, positive and negative. Franklin was one of the five men who drafted the Declaration of Independence in 1776. On his death, Turgot wrote Franklin's epitaph: 'He snatched lightning from the heavens and sceptres from kings.' [The Mansell Collection.]

gravity. Just how much stronger was well expressed by Richard Feynman in the second of his three famous 'red' books entitled *The Feynman Lectures on Physics*. Feynman describes the strength of electrical forces as follows:

> If you were standing at arm's length from someone and each of you had one per cent more electrons than protons, the repelling force would be incredible. How great? Enough to lift the Empire State Building? No! To lift Mount Everest? No! The repulsion would be enough to lift a 'weight' equal to that of the entire earth!

The forces that hold atoms together and the chemical forces that hold molecules together are actually electrical in origin. It is these enormous forces, perfectly balanced out by equal numbers of opposite charges, that give matter its strength. We are able to walk about on the surface of the Earth and feel the force of gravity precisely because of the almost complete screening of electrical forces.

Faraday had the idea that electrical forces exist in the space surrounding electric charges. It was this idea of the electric 'field' that ultimately led to special relativity. Changes in electrical forces resulting from movements of charges are not transmitted instantaneously. Such disturbances of the electric field travel at the speed of light. In fact, these disturbances *are* the electromagnetic waves that constitute light. In contrast, Newton's theory of gravity supposed that gravitational forces were transmitted instantaneously across empty space. Such 'action at a distance' seemed unnatural even to Newton, and it seems a small step to suppose that gravity, like electromagnetism, is a field theory. In such a theory, each massive particle or object would be surrounded by a gravitational field. Changes in the arrangements of these masses would cause a disturbance to propagate through the gravity field as gravity waves travelling with the speed of light. Such gravity waves could be generated, for example, by the motion of two stars in orbit about each other, or by something more violent, such as a 'supernova' explosion. Apart from differences of detail, this appears to be a very straightforward extrapolation of our ideas of electromagnetism. Unfortunately, there is a crucial difficulty concerned with the behaviour of light in a gravitational field.

So far in our analogy between electricity and gravity, it has been mass that played the role of a charge, acting as a source of the gravitational field. Electric charge is conserved: we cannot create or destroy electric charge in our theory. Before the discovery of radioactivity, it was believed that mass was also conserved – that mass could not be created or destroyed. In nuclear physics, using Einstein's famous mass–energy relation, conservation of *energy* is seen to be the fundamental law. Einstein showed that mass can be exchanged for a precise amount of energy. We must now ask the question: what is the source of the gravitational field – mass or energy? It seemed probable to Einstein that gravity was a force that acted on all energy, rather than on just one aspect of energy, namely the mass. This premise leads to the conclusion that light, being a form of energy, will experience the gravitational force. In terms of the theory of light as a stream of bullets, this is perhaps not

Figure 8.3 *Relativity* by the artist M. C. Escher. In the absence of gravity, the concepts of up and down are meaningless. [© 1995 M. C. Escher/Cordon Art – Baarn – Holland. All rights reserved.]

very surprising: in terms of the modern quantum wave–particle view of light, it is not obvious. Nor is it clear how special relativity, with its insistence on the constancy of the velocity of light, can be maintained if light is accelerated by gravity.

If light is affected by gravity, one direct consequence is the possibility of 'black holes'. Such objects – but not their name – were first suggested as long ago as 1793 by a British amateur astronomer, the Reverend John Michell. Some years later, the great French scientist and mathematician Pierre Simon Laplace had the same idea. Michell and Laplace both reasoned that light radiated by a sufficiently massive object would be unable to escape. Just like a stone thrown upwards from the Earth, the light 'particles' would be

164

slowed down and would fall back to the surface. Michell calculated that a star of the same density as our Sun would need to be 500 times its diameter for light not to be able to escape. Although wrong in details, both Michell and Laplace were correct in their basic ideas.

Einstein's suggestion that gravity acts on energy in all its forms, rather than on just mass, implies that a light beam will be 'bent' by gravity as it passes close to a massive body, such as the Sun. This seems at odds with the special theory of relativity and with the view that the speed of light is a universal constant of Nature. Gravity cannot be screened off, and gravitational forces are ever present throughout the universe. Should we give up special relativity as far as gravity is concerned? Or should we change the law of gravity so that light is unaffected, but so slow-moving bodies have weight in the usual manner? Such a theory was put forward by the Finnish physicist Gunnar Nordstrom in around 1912. Other physicists also tried to develop a theory of gravity without special relativity. One of these, Max Abraham, even argued that relativity was leading physics up a blind alley, since

> it was clear to the sober observer that this theory could never lead to a complete world picture if it were not possible … to incorporate gravity.

Einstein did not think much of these attempts, and commented privately that Abraham's theory was like 'a stately horse which lacked three legs'.

Figure 8.4 The Marquis Pierre Simon de Laplace (1749–1827) was expected to go into the Church after receiving an education in a Benedictine college. Instead, he went to Paris with a letter of recommendation to the scientist Jean d'Alembert, who led him into science. One of his first scientific appointments was as assistant to another famous French scientist, Lavoisier. Laplace became involved in politics and served as a minister and a senator – in addition to his important work in astronomy, mechanics and probability theory. He is best known for his five-volume book *Celestial Mechanics* which frequently contains the phrase 'it is obvious that', which is guaranteed to irritate the reader for whom 'it' is not at all obvious. Napoleon is said to have remarked: 'You have written this huge book on the system of the world without once mentioning the author of the universe.' To which Laplace is reputed to have replied: 'Sire, I have no need of that hypothesis.' [The Mansell Collection.]

How can special relativity be brought into peaceful co-existence with gravity? In 1907, Einstein had 'the happiest thought of my life'. This was the germ of the idea that enabled him to bring together special relativity and gravity. It was still a long hard road to his theory of general relativity, completed some eight years later, but this insight enabled Einstein to take the first step in the right direction. To appreciate Einstein's happy thought, we must remind ourselves of Galileo, Newton and their beliefs about mass and acceleration.

Figure 8.6 A portrait of Galileo Galilei (1564–1642) by Ottavio Leoni in 1624. At the time of this portrait, Galileo was aged sixty and was famous for his use of the telescope in astronomy. He was also deeply embroiled in his dispute with the Church. He had yet to write his two masterpieces *Dialogue on the Two Chief World Systems* and *Discourses Concerning Two New Sciences*. These two works contain most of his contributions to physics and establish him as one of the greatest scientists of all time. Galileo never married, but had three children during the period when he lived with a Venetian girl, Marina Gamba. [Paris, Musée du Louvre, Département des Arts Graphiques.]

Falling to Earth: Galileo and Eotvos

The gravitational field has only a relative existence… because for an observer falling freely from the roof of a house there exists… no gravitational field.

Albert Einstein, *Morgan Manuscript*, 1921

In 1969, millions of people around the world watched Neil Armstrong become the first man to step onto the surface of the Moon. In the lower gravity of the Moon, Armstrong and his co-astronaut, Buzz Aldrin, were able to gambol almost effortlessly on the lunar surface despite their bulky spacesuits. The Moon has only about one-eightieth of the mass of the Earth and has a

Figure 8.7 The Leaning Tower of Pisa. Galileo is supposed to have dropped two stone balls from the tower to show that all objects, regardless of their weight, will fall at the same rate.

Figure 8.5 Edwin (Buzz) Aldrin carrying heavy scientific equipment on the Moon. The low gravity made this task possible and enabled the astronauts to jump about, despite their bulky space suits. [NASA.]

Figure 8.8 The Apollo 15 astronaut David Scott performing a lunar version of Galileo's experiment. Scott dropped a hammer and a feather at the same instant, and the video record showed that they fell at exactly the same rate. [NASA.]

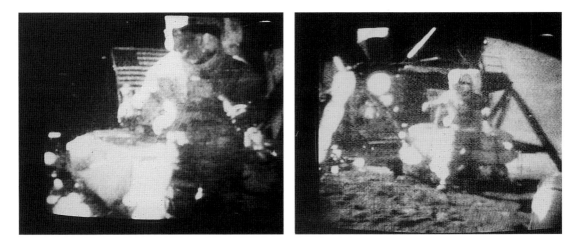

correspondingly weaker gravitational pull. This is partially compensated for, however, since the surface of the Moon is closer to the centre of gravitational attraction. The Moon's radius is about one-quarter that of the Earth, and so, by the 'inverse square' law of gravity (that the gravitational force experienced by an object is inversely proportional to the square of the distance between the object and the centre of attraction) the pull at the surface due to this distance effect is increased by a factor of 16 at the surface of the Moon relative to that on Earth. The net effect of both mass and distance means that, on the Moon, astronauts will experience about one-fifth of their weight on Earth. This is the standard, and correct, prediction from Newtonian gravity.

Of more concern to us here is the experiment carried out by *Apollo 15* astronaut David Scott. He repeated Galileo's famous experiment by dropping a hammer and a feather simultaneously onto the Moon. Observers watching television were able to see the feather and the hammer fall at precisely the same rate and reach the lunar surface at exactly the same time. Down on Earth, Scott's experiment would have had a very different result: the hammer would fall much faster than the feather because air resistance slows the feather much more than the hammer. The feather would waft gently to the floor long after the hammer. Other things on Earth further confuse the picture. Smoke particles, for example, show that some things rise rather than fall. The ancient Greek system of physics as formulated by Aristotle incorporated these common observations into an elaborate framework. In this system, 'gravity' was a quality that was possessed by bodies having weight: when

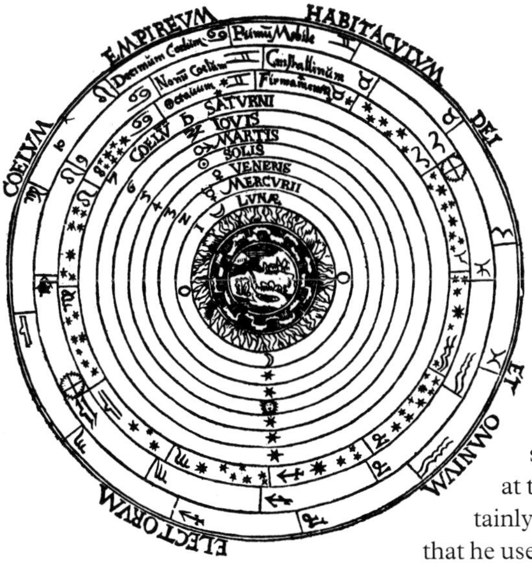

Figure 8.9 The Aristotelian system of the universe. This was a geo-centric system with the sphere of fire below the sphere of the Moon and heaven beyond the sphere of stars.

unsupported, such bodies fall, seeking the centre of the universe, which was also considered to be the centre of the Earth. Fire, on the other hand, possessed 'levity', which naturally moves upwards. Beyond the Earth, it was believed that the heavens were perfectly spherical and that the stars and planets moved in circular orbits. Aristotle believed that heavier bodies with more 'gravity' fall faster than less 'ponderous' objects. The influence of Aristotle over the 'establish-ment' in the following centuries made it difficult to challenge any of his pronouncements. In the case of gravity, Galileo may not have been the first to refute the idea that heavier objects fall faster, but he was certainly the most famous.

It was Galileo's biographer, Vivani, who started the legend that Galileo had dropped vari-ous objects from the Leaning Tower of Pisa. The purpose of these experiments was to demon-strate that all objects, regardless of their weight, fall at the same rate under the pull of gravity. Galileo cer-tainly performed many experiments, but it is doubtful that he used the famous Leaning Tower. He also considered a striking 'thought experiment', long before Einstein's famous 'gedanken experiments' of the twentieth century. Imagine that a building is ablaze, and that a man carrying an injured child jumps from the roof onto a stretched out blanket held by firemen below. If each body falls at a rate dependent upon their weight, the child will tend to fall slower than the man. The man will have to hold on tight to prevent the child rising above him as he falls. This suggests that the man and child will together fall at a rate part way between the rate of either. On the other hand, because the man and child together are heavier than either separately, according to Aristotle this combined bundle will fall together faster than either separately! This paradox shows that something must be wrong. Galileo concluded that, even if the man accidentally let go of the child, both would continue to fall at the same rate and thus continue to be at rest relative to each other.

In spite of this thought experiment, Galileo did not propose a full theory of gravity. This honour was left to Isaac Newton, who was born in 1642, the year of Galileo's death. One consequence of Newton's theory of gravity has particular relevance to Einstein's prediction of the bending of light: this was Newton's idea of artificial satellites. Newton arrived at this idea more than 300 years ago by the following reasoning. Imagine simultaneously dropping a stone from a height and firing a cannon from the same height. The cannon-ball starts out travelling horizontally, but, under gravity, falls to the Earth at exactly the same time as the stone that was dropped vertically. If we ignore details such as air resistance, we can imagine making the cannon more powerful and shooting the cannon-ball faster. Eventually, the cannon-ball will be moving at such a speed that, as it falls, the curvature of the Earth exactly compensates for the distance it has fallen. The cannon-ball is in orbit, just like the Moon. The Moon is further away than the cannon-ball and so 'falls' towards the Earth more slowly than the fast-moving cannon-ball. Light travels much faster than any bullet, but, if gravity acts on its energy just like it does on its matter equivalent, the light will fall by exactly the same amount as the cannon-ball in equal times. It was precisely this possibility that led Einstein to wonder how relativity could be adapted to include gravity.

In 1907, some 270 years after Galileo published his most famous work *Dialogues Concerning Two New Sciences*, Einstein was sitting in a chair in the patent office in Bern puzzling over these questions. He then had 'the happiest thought' of his life. He described this as follows:

> Because for an observer falling freely from the roof of a house there exists – at least in his immediate surroundings – no gravitational field. Indeed if the observer drops some bodies then these remain to him in a state of rest or uniform motion … . The observer therefore has the right to interpret his state as 'at rest'.

At least until he hits the ground! What Einstein had realized was that, in the freely falling 'weightless' state, gravity is abolished. Nowadays, the safest way to free-fall is in an orbiting space laboratory. Just like the cannon-ball, the laboratory falls freely towards the Earth by just the right amount to keep in orbit. Similarly, an observer in the freely falling laboratory would see that

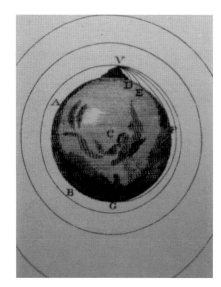

Figure 8.10 Newton's 'thought experiment' to explain the orbital motion of a very rapidly moving cannon ball. In a similar manner, the Moon may be viewed as falling round the Earth. This picture is taken from Newton's *A Treatise of the System of the World*, which was a more popular and less mathematical account of Newton's ideas about gravity, written in the 1680s but not published until after his death.

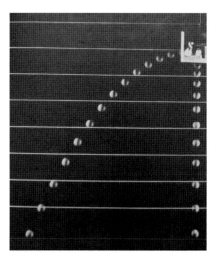

Figure 8.11 A stroboscopic photograph of the motion of a ball initially projected horizontally and a ball falling vertically. Both balls are released at the same time and fall at the same rate. From the point of view of someone falling with one of the balls, the other will be in uniform motion away from the ball he is sitting on.

light travels in a straight line: on Earth, an observer sees the laboratory, and the light beam, fall by the same amount. In such a laboratory, the observer would be able to check the laws of physics and special relativity in a situation with no gravity.

Just as gravity can be 'abolished' by free-fall, an artificial gravity can be created by acceleration. As a rocket lifts off from the launch pad, an astronaut feels heavier, and is pressed into the bunk by the so-called 'g-forces'. Similarly, if an aeroplane suddenly begins to descend, you feel pulled out of your seat. In the first case, some of the effect of gravity has been enhanced by acceleration; in the second case, some has been cancelled. Gravity and acceleration are, in a certain sense, equivalent: this is the origin of Einstein's new interpretation of Newton's 'equivalence principle'. To Newton, the 'equivalence principle' was the striking fact that the mass of a body that determined its response to a force – the 'inertia' that governs its rate of acceleration – appeared to be directly proportional to the weight of the body, which determines how strongly it reacts to a gravitational force. It is this coincidence that ensures that all bodies fall at the same rate in a gravitational field, regardless of their mass. Newton was not content to accept such a coincidence as fact without performing some experimental tests. He constructed a pair of identical pendulums consisting of a wooden box at the end of an 11 foot wire. One box was then filled with wood, and the other was filled with the same weight of gold. If Newton's 'equivalence' of 'inertial' mass and gravitational weight was correct, the period of the swings of the pendulums should depend only on the length of the wire suspending the boxes. When set in motion in identi-

Figure 8.12 A free-floating sphere of water produced during Skylab simulation training in 1972. [NASA.]

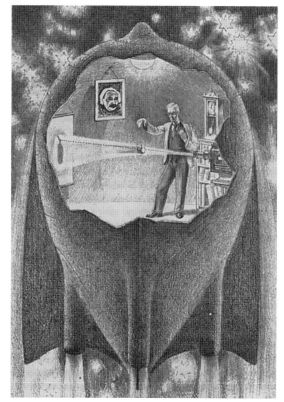

Figure 8.13 An illustration of Einstein's principle of equivalence taken from the book *Mr Tompkins in Wonderland* by George Gamow. For someone inside the spacecraft, it is impossible to distinguish acceleration of the rocket from the gravitational attraction of a nearby object. The original caption was 'The floor... will finally overtake the apple and hit it.' [Courtesy of Elfriede Gamow.]

cal fashion, the pendulum pairs kept in step, regardless of what material was in the boxes. Newton concluded that the mass and weight of materials are the same to an accuracy of at least 1 part in 1000.

It was five years after Einstein had proposed his equivalence principle that he learned of some experiments, performed towards the end of the nineteenth century, which significantly improved upon Newton's accuracy. The Hungarian physicist Roland von Eotvos had developed a device called a 'torsion balance'. This consisted of a pair of objects of equal mass attached to the opposite ends of a rod suspended by a fine wire. There is, of course, a gravitational force on each object, and, in addition, an inertial force due to the rotation of the Earth. If the gravitational and inertial masses are not exactly equal, the rod will rotate about a vertical axis until it is halted by the restoring torque of the twisted wire. The amount of twist is best measured by turning the whole apparatus, including the wire and its support, through 180 degrees. The rod will now twist in the opposite direction. If the two gravitational and inertial masses are equal, there will be no movement of the rod between the two orientations of the apparatus. Using this apparatus, Eotvos was able to demonstrate their equality to within a few parts in 1000 million.

In the 1960s, a group, led by Robert Dicke from Princeton University in the USA, were able to improve on this result of Eotvos. Some ten years later, together with a Russian group headed by Vladimir Braginsky at Moscow State University, they were able to make a further improvement on the equality of the two types of mass. The accuracy of the equality now stands at about 1 part in 100 000 million. Less accurate, but still significant, experiments have also shown that molecules, neutrons and electrons also fall exactly in step with dust particles and other objects. It is also interesting that technology has now improved sufficiently for the direct test of dropping two objects and comparing their descent to be as accurate as the more indirect Eotvos experiment one hundred years earlier. This new generation of experiments was motivated by the suggestion that gravity had an additional short range force component that played no role in the large-scale astronomical manifestations of gravity. A variety of detailed experiments carried out between 1986 and 1990 now seem to rule out the existence of such an additional force.

Gravity, time and red-shifts

Gravity is probably the most familiar of forces. We are acutely aware of it when climbing mountains, or even going up stairs. Yet we are no longer surprised to see satellite pictures of astronauts floating in their space laboratories. Everything in the orbiting laboratory seems to be floating – from drops of water to bottles of aftershave – since all objects are falling towards Earth with the same acceleration. It certainly looks as if gravity has been abolished. In fact, this is only approximately true. Imagine specks of dust sprinkled throughout the spacecraft. Those specks nearest the Earth will be pulled towards the Earth slightly more strongly than those further away. Since the spacecraft is solid, it is falling towards the Earth at an intermediate rate. Inside the craft, the dust on the side nearest the Earth will fall towards the 'Earthside' of the laboratory. Conversely, dust in the half of the ship furthest from the Earth will be seen to rise and settle on the 'topside' of the laboratory. In the case of 'spaceship' Earth in its orbit round the Sun, a similar effect

Figure 8.14 A Guinness advertisement.

Figure 8.15 The motion of a ray of light in an accelerated rocket suggests that light is deflected by gravity. While true, this effect only explains part of the bending of star light by the Sun. [Courtesy Elfriede Gamow.]

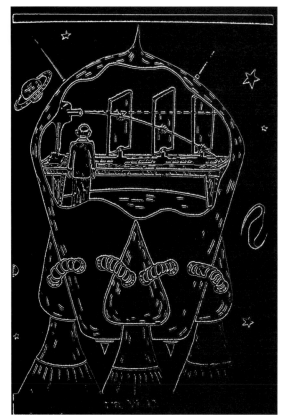

takes place with the air and water, and this is the origin of the solar tides. The much larger lunar tides arise in the same way, from the effects resulting from the Moon orbiting the Earth. These residual effects of gravity in our freely falling laboratory are called 'tidal forces': as we make the laboratory smaller, the smaller these tidal forces become and the closer we are to a true gravity-free situation. To describe fully the magnitude of such tidal forces requires a full theory of gravity. Before we describe Einstein's theory of gravity, we shall describe some predictions that follow from the equivalence principle alone.

We have talked about the bending of light by gravity, and historically this was the first important test for Einstein's new theory. However, although the equivalence principle does predict the bending of light, the full apparatus of general relativity is required to predict the magnitude of this deflection. Figure 8.15 shows an accelerating rocket with a laser beam of light. In the absence of acceleration, the light will travel in a straight line; with the rocket accelerating upwards, the light is seen to bend. According to the equivalence principle, light must also be bent by a gravitational field. Einstein first attempted to calculate the magnitude of the bending of light by the Sun in 1911. In fact, although he did not know it at the time, he had obtained the same answer as one first obtained in 1803 by a little-known Bavarian

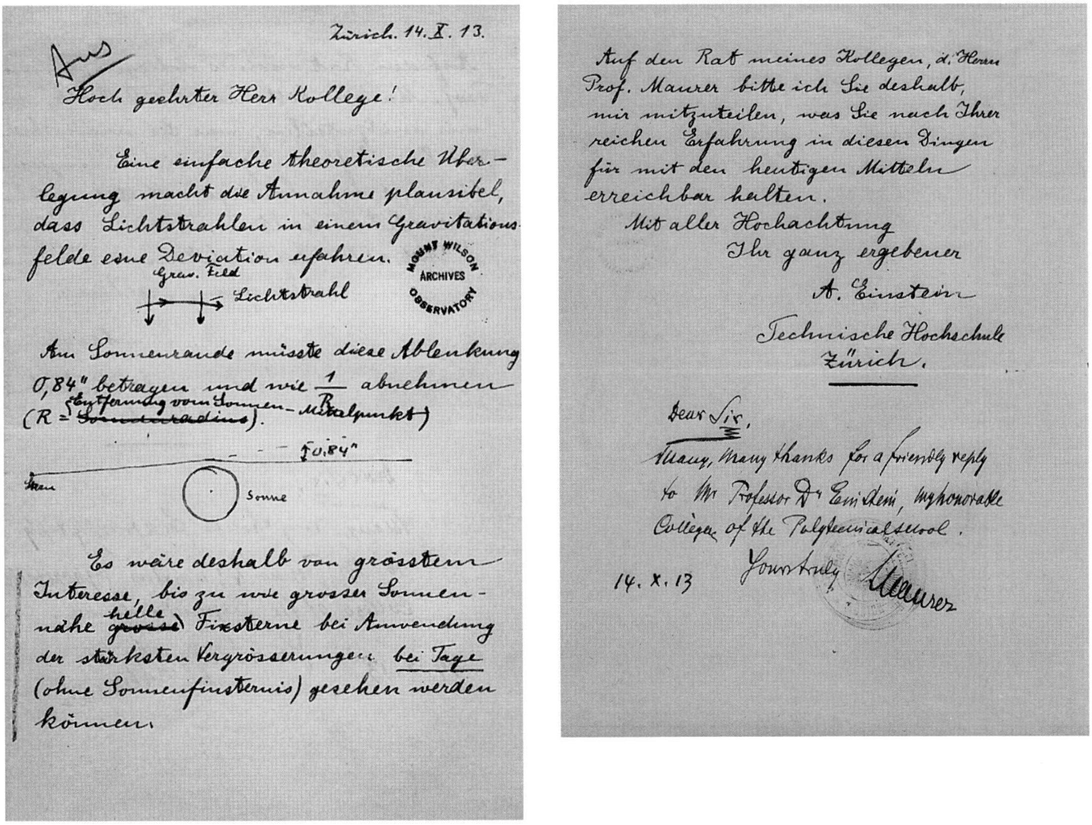

Zürich. 14. X. 13.

Aus

Hoch geehrter Herr Kollege!

Eine einfache theoretische Überlegung macht die Annahme plausibel, dass Lichtstrahlen in einem Gravitationsfelde eine Deviation erfahren.

Grav. Feld

Lichtstrahl

Am Sonnenrande müsste diese Ablenkung 0,84" betragen und wie $\frac{1}{R}$ abnehmen (R = ~~Sonnenradius~~ Entfernung vom Sonnen-Mittelpunkt).

0,84"

Sonne

Es wäre deshalb von grösstem Interesse, bis zu wie grosser Sonnennähe helle ~~grosse~~ Fixsterne bei Anwendung der stärksten Vergrösserungen bei Tage (ohne Sonnenfinsternis) gesehen werden können.

MOUNT WILSON OBSERVATORY ARCHIVES

Auf den Rat meines Kollegen, d. Herrn Prof. Maurer bitte ich Sie deshalb, mir mitzuteilen, was Sie nach Ihrer reichen Erfahrung in diesen Dingen für mit den heutigen Mitteln erreichbar halten.

Mit aller Hochachtung

Ihr ganz ergebener

A. Einstein

Technische Hochschule Zürich.

Dear Sir,

Many, many thanks for a friendly reply to Mr Professor Dr Einstein, my honorable College of the Polytechnicalschool.

14. X. 13 Yours truly Maurer

Figure 8.16 A copy of the letter Einstein wrote to George Hale about the possibility of detecting the deflection of star light by the gravity of the Sun. [Reproduced by permission of The Huntington Library, San Marino, California.]

astronomer Johann Georg van Soldner, who used Newtonian gravitational theory and treated light as a stream of bullets. The effect was found to be very small, and in 1913 Einstein wrote to the American astronomer George Hale to ask if it were possible to test this prediction without waiting for an eclipse. Hale replied that an eclipse was necessary: stars near the Sun would then be visible, and the bending of light from these stars would show up as an apparent displacement of the stars from their normal positions. The German astronomer Erwin Finley-Freundlich planned an expedition to Russia to observe an eclipse due to occur there in 1914 and thus to test Einstein's prediction. Fortunately for Einstein, the First World War intervened, and no observations could be made. Why fortunately? Because, by 1915, Einstein had completed his general theory of gravitation and he had repeated his calculation. He found that the deflection should be exactly twice as large as he had originally predicted!

The most important experimental consequences of the equivalence principle are the gravitational red-shift and the related slowing down of time by a gravitational field. Einstein devised a thought experiment to illustrate this effect. Imagine a continuous belt running vertically around wheels at the top and bottom of a tower on Earth (Figure 8.17). In this idealized experi-

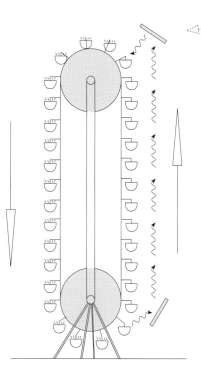

Rotating mirror

Figure 8.17 A 'thought experiment' to illustrate the existence of the gravitational red-shift. Light is transferred from the atoms at the bottom of the tower to those at the top by mirrors. Without the gravitational red-shift we would be able to make a perpetual motion machine. The red-shift may be compensated for by the Doppler shift generated by a rotating mirror.

ment, we can ignore friction and suppose the belt to be manufactured from a material whose atoms can be made to emit or absorb light of a specific frequency whenever required. We arrange things so that only atoms on one side of the belt are in an excited atomic state. These atoms have more energy than their unexcited counterparts, and therefore have more mass and are heavier. This imbalance of heavier atoms on one side causes the belt to rotate, the side with the excited atoms moving downwards. Just as the excited atoms reach the bottom, they emit their extra energy as light, which is then directed by a system of mirrors up to the top of the belt. As the unexcited atoms reach the top, they are able to absorb this light energy and become excited in time to start their journey down again. We seem to have created a perpetual motion machine which will run for ever and from which we could extract useful work! Such a machine must be impossible. The problem lies not in our idealizations but in a key idea that we have overlooked. The missing factor is the work that must be done against gravity to raise the light energy – equivalent to mass – from the bottom to the top of the tower. This energy loss corresponds to a shift in observed frequency of the light towards the red – the gravitational red-shift.

We could make our idealized apparatus work by adding a rotating mirror at the top of the tower. The mirror acts as a moving source of light and, according to Doppler, alters the observed frequency of the light. The speed of rotation can then be adjusted so that the Doppler shift imparted to the reflected light exactly compensates the energy lost in travelling up the tower against gravity. The rotating mirror does work against the pressure of the incoming light and imparts a compensating blue-shift to the light. In the best case, the energy gained from the rotating belt will just be sufficient to drive the compensating mirror, but not enough to do useful work.

Einstein concluded that light moving against a gravitational field will be red-shifted. This conclusion will be true of almost any realistic theory of gravity. This gravitational red-shift also implies that time is slowed by a gravitational field. In atomic clocks, the measurement of time is by the frequency of microwave vibrations. When viewed from the top of the tower, the clock at the bottom will be seen to be running slower than an identical clock at the top. Similarly, light falling from the top of the tower will gain energy and be blue-shifted: an observer at the bottom will therefore see the clock at the top of the tower running faster. Both observers agree that time, as measured by these clocks, runs slower in a higher gravitational field. In our discussion of the GPS positioning system in chapter 1 we had to correct for this effect – which was in the opposite direction to the time dilation caused by the speed of the satellite. An extreme case of this red-shift and slowing of time is provided by a black hole. Anything falling past the 'event horizon' of a black hole can never escape its gravitational pull. For an external observer, any light emitted by an object falling into the black hole would appear more and more red-shifted as it approached the event horizon. Clocks carried by the object would appear to run slower and slower as the object approached the black hole. When the object reached the event horizon,

Life on a neutron star

Following a supernova explosion, it is believed that either a neutron star or a black hole is left behind. For a star remnant about as massive as our Sun, the resulting neutron star would be only about 10 miles in diameter. The density of neutron star matter is predicted to be more than one hundred million million times that of water and about the same as that in an atomic nucleus. From the 'reality' school of science fiction comes the novel *Dragon's Egg*, by Robert L. Forward, which is about an intelligent civilization living on the surface of a neutron star. Although this is an entertaining idea, life on a neutron star would be almost two dimensional, since the strength of the gravitational field restricts the highest mountains to only a few inches in height, and consequently room for fiction is rather limited. If we drop such irritating restrictions and move into a fantasy world, we can imagine some striking effects that would be caused by the enormous gravitational field. Suppose our fantasy life-forms on the neutron star work in tower office blocks. If the lighting on the ground floor is red, then, looking upwards, we would see the lighting on the upper floors progressively shift to the blue end of the spectrum. Working on the top floor, life for the office workers would seem normal, apart from the lower gravity compared with that on the ground. Light coming straight out from the lower floor windows would be bent, falling to the surface of the star under the huge gravitational pull. Just like Newton's cannonballs, the degree of bending gradually becomes less until workers in the higher floors can see right round the star and look in at the window on the other side of the building. When office workers return from a day's work on the top floor, they discover that only half a day has passed on the ground. Persistent high-floor workers would live shorter lives than the ground-floor workers, although they would accomplish the same amount of work in a lifetime. This is the gravitational equivalent of the famous twin paradox of special relativity. [*Dragon's Egg* by Robert L. Forward (New English Library, 1981).]

the red-shift of the light emitted would become infinite, time would effectively stop, and the object would disappear from view.

How can we observe Einstein's predicted red-shift? In his original paper, Einstein suggested that this gravitational red-shift could be observed in the light from stars. In practice, this observation has proved rather difficult. In the case of light emitted by atoms at the surface of the Sun, the atoms are undergoing violent motions due to rising and falling columns of gas leading to both red and blue Doppler shifts. It was only in the late 1960s, by which time these surface effects were better understood, that any sort of test of the gravitational contribution to the red-shift was possible. In 1971, Joseph Snider used solar spectra to show that Einstein's prediction for the gravitational red-shift agreed with data to within about 5% accuracy. Other attempts to measure this red-shift in stars have centred on white dwarfs. These are the collapsed remains of burnt out stars with roughly the same mass as the Sun, but which are contained in a volume around the same as that of the Earth. The much larger gravitational force at the surface of these stars leads to predicted red-shifts one hundred times larger than those from the Sun. Unfortunately, it is necessary to know the mass and radius of the white dwarf accurately to make an accurate prediction of the expected red-shift;

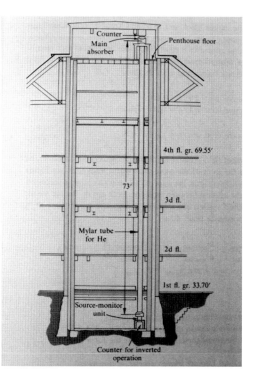

Figure 8.18 Diagram of the arrangement of apparatus in the Harvard physics tower which measured Einstein's gravitational red-shift. Gamma rays – very high energy photons – travelled up the tower in a helium filled tube.

and, consequently, such white dwarf stars have not proved useful as a test of Einstein's gravitational red-shift. Surprisingly perhaps, the most accurate tests of Einstein's red-shift predictions have come from experiments based on Earth.

The terrestrial test of the gravitational red-shift almost mimics Einstein's thought experiment by which he established the existence of the effect. It is concerned with measuring the red-shift produced by light travelling from the bottom to the top of a tower. The tower used in practice was the Jefferson Tower of the physics building at Harvard University, and the experimentalists were Robert V. Pound and Glen A. Rebka, Jr. The challenge of an experiment in a tower only 74 feet high is that the predicted frequency shift is tiny – only about 2 parts in 1000 million million. The problem is that, as light is emitted by an excited atom, to conserve momentum the atom must recoil in the opposite direction. This recoil causes a Doppler shift, which broadens the frequency of the emitted light so much that any test to measure such a small predicted red-shift is made impossible.

The Pound–Rebka experiment was only made possible by a phenomenon discovered by a German physicist, Rudolf Ludwig Mössbauer. From 1955 to 1957, Mössbauer was doing research for his doctorate at the Max Planck Institute for Medical Research in Heidelberg. He discovered that he could eliminate most of this recoil Doppler effect for photons emitted from a radioactive nucleus that was bound in a crystal. Mössbauer originally used an isotope of iridium, but a similar effect is obtained with a radioactive iron nucleus. The nucleus of radioactive cobalt 57 can capture an electron and transmute into an excited nucleus of iron 57 (Fe57). This iron isotope is very unstable, and rapidly emits a gamma ray photon with a very precise frequency. When an iron nucleus decays or absorbs such a photon, it must recoil to ensure conservation of momentum. What Mössbauer discovered was that if such a nucleus is embedded in certain types of crystal, then the forces of the surrounding atoms transfer the troublesome recoil momentum to the entire crystal – rather than to the individual iron atom. The huge mass of the crystal will only move very slowly to balance the momentum, and so the Doppler shift is almost eliminated. Mössbauer received the Nobel Prize for physics in 1961, one year after Pound and Rebka's famous experiment.

Pound and Rebka placed the Fe57 gamma ray source at the bottom of the Jefferson Tower on a hydraulically movable platform. The Fe57 absorber at the top only absorbed the gamma rays strongly at their original frequency. In order to measure the effect of the red-shift, Pound and Rebka raised the platform slowly as the source emitted its gamma rays. This produced a small Doppler shift towards the blue end of the spectrum that compensated for the gravitational red-shift. The red-shift can therefore be determined by measuring the Doppler shift required to produce maximum absorption. The required velocity was about 2 millimetres/hour. In order to eliminate possible errors, the experiment was repeated with the source at the top of the tower and the absorber at the bottom. Pound and Rebka's paper, entitled 'Apparent weight of photons', was published in the April 1st issue of *Physical Review Letters* in 1960, and their data verified the gravitational red-

Figure 8.19 Rudolf Ludwig Mössbauer was a graduate student in 1958 when he made the discovery that was to win him the Nobel Prize in 1961. His experiments showed that most of the recoil Doppler effect was eliminated for gamma decays of excited states of radioactive nuclei bound in a crystal. The photon can then be absorbed by another nucleus at rest in its ground state. [Courtesy Prof. R. L. Mössbauer.]

Figure 8.20 On the left, G. A. Rebka with the source
of gamma radiation at the bottom of the tower; on
the right, R. V. Pound at the top of the tower with
the absorber, counter and related apparatus.
[Courtesy R. V. Pound.]

shift to within about 10%. An improved version of this experiment, performed by Pound with Joseph Snider, obtained agreement with Einstein's
prediction to 1%.

In this chapter, we have described Einstein's first attempt to
formulate a consistent theory of gravity. His 'happiest thought' led to the
equivalence principle, verified in the Eotvos and Dicke experiments, and
the gravitational red-shift, measured in the Jefferson Tower experiments. In
between this thought in 1907, and his unveiling of his general theory of
relativity in 1915, lay a transformation in Einstein's thinking about the world.
It is to the world of curved space-time that we must now turn our attention.

9   Warped space

Geometry and gravity

The discovery of 'non-Euclidean' geometry in the nineteenth century came
as a great surprise and was greeted by disbelief. One of the pioneers of this
new geometry, Janos Bolyai, a Hungarian army officer, expressed his joy with
the words:

> I have made such wonderful discoveries that I am myself lost in astonish-
> ment. Out of nothing I have created a new and another world.

'Euclidean' geometry is the geometry we learn in school, with its familiar
apparatus of points, straight lines, circles, ellipses and triangles. In particular,
we are all brought up to believe that the three angles of a triangle add up to
180 degrees and that parallel lines never meet. Such Euclidean geometry is
the geometry of the plane – technically called a 'flat' space. By contrast, non-
Euclidean geometry describes a 'curved' space. What do we mean by these
terms?

Some idea of a curved space can be gained by considering geometry
on the surface of the Earth. The Earth is approximately spherical, and on its
surface it is easy to construct triangles whose angles add up to more than

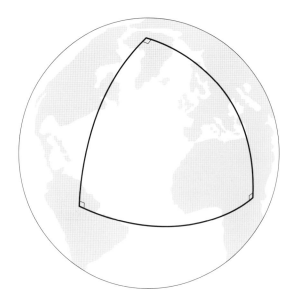

Figure 9.1 A triangle on the surface of the Earth for which the sum of the three interior angles adds up to more than 180 degrees.

180 degrees (Figure 9.1). Similarly, lines of longitude start out parallel at the equator but converge and cross at the poles. The surface as a whole does not obey Euclid's rules. Since such a familiar example of a surface is non-Euclidean, why are such geometries so unfamiliar to most of us? If we restrict ourselves to a very small portion of the Earth's surface – such as an area the size of the page of this book – then the curvature of the Earth is negligibly small and we recover the familiar geometry of Euclid. For most purposes, the fact that we live on the surface of a sphere does not reveal itself.

What has this new geometry got to do with physics? The great French mathematician, Henri Poincaré, wrote:

> One geometry cannot be more true than another; it can only be more convenient. Now Euclidean geometry is, and will remain, the most convenient.

Einstein thought otherwise: non-Euclidean geometry was the basis for his new theory of gravity. His equivalence principle had solved one problem concerning gravity but had raised another. We have seen how small, freely falling regions of space are effectively gravity-free, and that special relativity and the other laws of physics are valid inside this region. The problem is how to link these small, free-fall regions together. Just as each small region on the Earth's surface looks like a flat Euclidean space, when we link them all together we find that the space is curved. For a small portion of the Earth's surface, a flat map is adequate, but, as we increase the area covered by the map, distortions become apparent. For example, maps which give the correct size of regions round the equator are well known to give great

Figure 9.2 Any attempt to represent the surface of a sphere on a flat surface will always result in some distortion of distances and size. In the usual 'Mercator projection' of the Earth's surface, there is an inevitable east–west stretching since all lines parallel to the equator are assumed to have the same circumference as the equator. In order to show a constant compass direction as a straight line on the map, Mercator introduced a compensating north–south stretching: this makes the scale along the meridians at any point equal to the scale of the corresponding parallels. This leads to reasonably accurate representations of the shapes of small areas, but causes considerable distortions of large areas the further north or south one goes. For example, South America is about ten times the size of Greenland, but on a Mercator projection Greenland appears larger. Mercator's projection is useful because it allows navigators to plot lines of constant compass bearings as straight lines – called rhumb lines. But only rhumb lines that run north–south or east–west along the equator can be part of great circles and correspond to the shortest distance route. The figure shows both the rhumb line and great circle routes between Cape Town and New York. The shorter great circle route appears as a longer curved path in this projection.

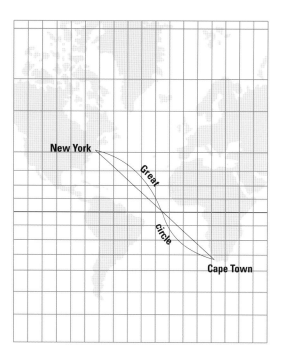

distortions of the polar regions. How then are these free-fall regions patched together to make a theory of gravity? If the size of these regions is increased, gravitational tidal forces become apparent. What is needed is some way of describing the effects of gravity over a large region. Einstein's answer was curved 'space-time'.

Einstein had tremendous difficulties in trying to turn this idea into a fully fledged theory. He described his position at the time with these words:

> If all [accelerated] systems are equivalent, then Euclidean geometry cannot hold in all of them. To throw out geometry and keep [physical] laws is equivalent to describing thoughts without words. We must search for words before we can express thoughts. What must we search for at this point? The problem remained insoluble to me until 1912, when I suddenly realized that Gauss's theory of surfaces holds the key for unlocking this mystery. I realized that the foundations of geometry have a physical significance. My dear friend the mathematician Grossmann was there when I returned from Prague to Zurich. From him I learned for the first time about Ricci and later about Riemann. So I asked my friend whether my problem could be solved by Riemann's theory.

During this time in Zurich, Einstein wrote to a colleague:

> I am now exclusively occupied with the problem of gravitation, and hope, with the help of a local mathematician friend, to overcome all the difficulties. One thing is certain, however, that never in my life have I been quite so tormented.

Figure 9.3 Marcel Grossmann (1878–1936) helped Einstein at several crucial points in his life: as a student, Grossmann lent Einstein notes to lectures which Einstein had skipped; after graduation, Grossmann helped Einstein obtain his first job in the patent office. After Einstein's success with special relativity, Grossmann guided Einstein through the mathematical difficulties of curved space and helped him formulate general relativity. In 1955, Einstein wrote: 'The need to express at least once in my life my gratitude to Marcel Grossmann gave me the courage to write this…autobiographical sketch.'

A great respect for mathematics has been instilled within me, the subtler aspects of which, in my stupidity, I regarded until now as pure luxury. Against this problem, the original problem of the theory of relativity is child's play.

In order to appreciate something of Einstein's radical new approach to gravity, we must first say something about non-Euclidean geometry.

Non-Euclidean geometry arose from a problem within Euclid's geometry. In about 300 BC, Euclid of Alexandria wrote the *Elements* – thirteen books systematizing geometry. The curious name derives from Plato's belief about a relationship between the Greek 'elements' – earth, fire, air and water – and the regular solids – cube, tetrahedron, octahedron and icosahedron. (The fifth regular solid, the dodecahedron, was identified with the aether.) The approach to geometry laid down by Euclid has formed the basis for teaching the subject for more than 2000 years, perhaps to the regret of many generations of schoolchildren. Euclid's geometry is based on five postulates, or 'axioms', which were assumed to be self-evidently true. However, the fifth axiom, or so-called 'parallel postulate', was not so obvious. This can be stated in many equivalent ways, and in its original form the word parallel was not mentioned. Nowadays, the standard form of this postulate is known as the 'Playfair axiom', after the British mathematician John Playfair. In this form, the fifth postulate states that, given a line, and a point not on this line, only one line through this point does not intersect the original line: this is the line parallel to the first line. Over the years, mathematicians began to think that this axiom was not obvious, and some doubted its validity.

Towards the end of the eighteenth century, Immanuel Kant, a great German philosopher and scientist, declared that Euclid's geometry was true, independent of experience. Despite such an intimidating assertion, in 1817 Karl Friedrich Gauss wrote:

Figure 9.4 *Print Gallery* by M. C. Escher gives a hint of the possibilities of non-Euclidean geometries. [© 1996 M. C. Escher/Cordon Art – Baarn – Holland. All rights reserved.]

Figure 9.5 *Circle Limit III* by
M. C. Escher depicts Lobachevsky
geometry – each fish must be
thought of as being of the same
size. Lobachevsky geometry
covers infinite space, so, in this
representation, due to Poincaré,
the fish appear very crowded
near the boundary. [© 1996 M. C.
Escher/Cordon Art – Baarn – Holland.
All rights reserved.]

> I became more and more convinced that the necessity of our [Euclidean]
> geometry cannot be demonstrated … we must consider geometry as of equal
> rank, not with arithmetic, which is purely logical, but with mechanics, which
> is empirical.

In other words, Gauss was saying that the true geometry of space must be
determined by experiment – not postulated as axioms. Gauss was a famous
mathematician, astronomer and physicist, but he was also an engineer who
practiced surveying. When asked in 1820, by his new patron, King George IV
of the United Kingdom and of Hanover, to survey the Kingdom of Hanover,
he is reported to have remarked that it was much the same whether he measured the position of a star or a church tower since both are applications of
the same geometry. Gauss actually performed the experiment of measuring
the angles of a triangle with sides over a 100 kilometres long. Although he
found no significant deviation from Euclid, Gauss continued to work on the
problem, keeping most of his heretical doubts to himself. In a letter in 1824
he wrote:

Indeed I have from time to time in jest expressed the desire that Euclidean geometry would not be correct.

Non-Euclidean geometry was discovered independently, and more or less simultaneously, by three men: a German, Gauss; a Hungarian, Janos Bolyai; and a Russian, Nicolai Ivanovich Lobachevsky. In 1823, Lobachevsky suddenly had the inspiration that a new geometry could be built without Euclid's fifth axiom. In Lobachevsky's geometry, first announced in 1826 and published in 1829, the sum of the angles of a triangle is less than 180 degrees. Gauss had also discovered an example of non-Euclidean geometry, but had not published it for fear of 'the clamour of the Boeotians'. This was a reference to the people from a region of ancient Greece famous for their obtuseness. When, in 1840, Gauss became aware of Lobachevsky's work, he engineered the Russian's election to the Göttingen Academy, which Lobachevsky entered two years later.

Wolfgang Bolyai was a Hungarian friend of Gauss who had spent a large part of his life trying to prove Euclid's fifth axiom. When he learnt that his son Janos was following the same quest, he wrote to him in an effort to dissuade him:

> For God's sake, I beseech you, give it up. Fear it no less than sensual passions because it, too, may take all your time, and deprive you of your health, peace of mind and happiness in life.

Even such strong words as these failed to persuade Janos to give up his first love – the search for non-Euclidean geometries. In 1829, Janos sent word of his conclusions to his father. His father then wrote to Gauss asking for his opinion on the unorthodox views of his son. Gauss replied that he was unable to praise Janos's work for it would mean self-praise, as he himself had held similar views for several years. Wolfgang Bolyai published his son's work in 1831 as an appendix to one of his own books. The subject of non-Euclidean geometry received little attention until Gauss's correspondence on the subject was published after his death in 1855.

The next important step was taken by Bernhard Riemann, who rescued non-Euclidean geometry from the forgotten fringes of mathematics and brought the subject into the mainstream. As a young man studying to be a clergyman, Riemann began to doubt another of Euclid's axioms: namely, that a straight line can be extended indefinitely in either direction. Riemann persuaded his father to allow him to change his career and to study mathematics under Gauss in Göttingen. Riemann proposed that lines can be finite in length but endless. This can be visualized if we look at the example of the equator on the Earth and interpret this as a 'straight line', a concept that is still consistent with the definition of a straight line as the shortest distance between two points. There is nothing special about the equator – any straight line between two points on the surface of a sphere is part of a 'great circle', defined such that a slice through a great circle cuts the sphere into two equal halves. This generalization of the notion of a straight line is called a 'geodesic'

Figure 9.6 Nikolai Ivanovich Lobachevsky (1792–1856) was born in Kasan, Russia, to poor parents. In 1800, his mother enrolled him in the local high school, which, in 1805, became the core of the new University of Kasan. After a turbulent period in Kasan, Lobachevsky was elected Rector of the University in 1826. Apart from the routine management of new appointments and buildings, Lobachevsky also had to organize the response to a cholera outbreak in 1830. A small booklet explaining Lobachevsky's new geometry was published in German in 1840. Gauss was sent a copy, and then learned Russian in order to read Lobachevsky's earlier papers.

Figure 9.7 Karl Friedrich Gauss (1777–1855) was one of the greatest mathematicians of all time. He was a child prodigy with an extraordinary ability for mental arithmetic. At the age of fourteen, he became a protegee of the Duke of Brunswick, who paid for his education. Gauss grew up to become Director of the Observatory at Göttingen, where he was able to pursue his research with few distractions. Gauss was something of a recluse and, like Newton, was reluctant to publish anything unless his results were complete. Besides declining to publish his work on non-Euclidean geometry, Gauss also failed to publish other pioneering work on complex numbers and Newtonian dynamics. [Photo AKG London.]

– Earth divider. The geodesics are paths of shortest length and define the straight lines for a given surface. In Riemann's novel geometry there are no parallel lines and the angles of a triangle add up to more than 180 degrees. A geometry with these properties is now known as a 'Riemannian', or elliptic, geometry. Mathematicians now recognize two types of non-Euclidean geometries. The other geometry is called the previously mentioned 'Lobachevskian', or hyperbolic geometry, in which the sum of the angles of a triangle add up to less than 180 degrees and there is more than one line through a point which is parallel to any given line.

In 1853, when Gauss was seventy-six, his star pupil, Riemann, was required to give a lecture to the Faculty at Göttingen in order to confirm his position as a lecturer. The tradition was that the would-be lecturer submitted a choice of three possible topics for the lecture, but the choice would be made between only the first two. For this reason Riemann had not fully prepared his third topic, the foundations of geometry. To Gauss, the prospect of hearing his best student lecture on the subject he had struggled with for much of his life was irresistible: he broke with tradition and chose the third topic. After several postponements, Riemann eventually delivered his lecture: 'On the hypotheses which lie at the foundation of geometry' in June 1854. At the end of the lecture, Gauss was reportedly uncharacteristically enthusiastic. What Riemann had done in his lecture was to extend Gauss's results on the curvature of two-dimensional surfaces to spaces of any number of dimensions – which cannot easily be visualized. In Euclidean geometry, the distance between any two points can be calculated in terms of their 'Cartesian' coordinates – their x and y values – using the theorem of Pythagoras. Now consider

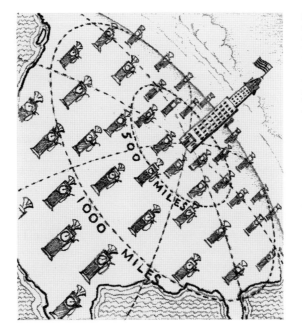

Figure 9.8 An illustration from the adventures of Mr Tompkins, which attempts to show a uniform distribution of filling stations across the United States, as measured from Kansas City. On the curved surface of the Earth, the number of filling stations will increase more slowly with distance from Kansas than if the Earth's surface were flat. The owner of the filling stations, Mr Shell, is wrong in thinking that his managers had located a higher concentration of stations near Kansas City. '[The professor said gravely] "He has forgotten that the earth's surface is not a plane but a sphere. And on a sphere the area within a given radius grows more slowly with the radius than on a plane."' [Courtesy Elfriede Gamow.]

Figure 9.9 Georg Friedrich Bernhard Riemann (1826–1866) had made many significant discoveries in mathematics before he died of tuberculosis at the age of forty. About Riemann, Einstein was moved to write the following epitaph: 'But…physicists were still far removed from such a way of thinking; space was still, for them, a rigid, homogeneous something, susceptible of no change or conditions. Only the genius of Riemann, solitary and uncomprehending, had already won its way by the middle of the last century to a new conception of space, in which space was deprived of its rigidity, and in which its power to take part in physical events was recognised as possible.'

how we calculate the distance between two points on the surface of a sphere. The position of a point on a globe can be specified in terms of latitude and longitude. The distance between even two quite close points on a sphere will be given by a complicated formula involving more than just the differences in latitude and longitude of the two points. It is complicated because the distance corresponding to a difference of 1 degree of longitude depends on where you are on the surface. A difference of 1 degree of longitude at the equator corresponds to a distance of about 110 kilometres (60 nautical miles), but this distance decreases as we move north or south in latitude, until it becomes zero at the poles. Riemann discovered that distances on any surface can be calculated by a generalization of the Pythagoras theorem that involves some new quantities called the 'metric' of the surface. Riemann extended this geometry to more than two dimensions and to surfaces whose curvature was able to vary from place to place. This was exactly the mathematical apparatus that Einstein needed to construct his new theory of gravity.

Einstein connected the equivalence principle with curved space-time. The equivalence principle predicts that gravity slows down clocks. In other words, a change of one unit of time at the bottom of the Jefferson Tower is different from one at the top. This is very like the difference of a unit of longitude on the curved space of our Earth. The message Einstein took from this was that the geometry of space-time in the presence of gravity is curved.

General relativity

…the happy achievement seems almost a matter of course, and any intelligent student can grasp it without too much trouble. But the years of anxious searching in the dark, with their intense longing, their alternations of confidence and exhaustion and the final emergence into light – only those who have experienced it can understand that.
Albert Einstein, *Notes on the Origin of the General Theory of Relativity*, 1934

What is general relativity? It is a theory of gravity in which Einstein assumes not only that the effects of gravity can be described in terms of a curved space-time but goes further and obtains the relationship between the curvature of space-time and the distribution of matter. In the presence of a gravitational field, the geometry of space-time, as described by the metric, is curved. There are two questions to be answered. (1) Given the distribution of matter, what is the space-time metric? Then, (2) given the metric, how does matter move? After an eight-year struggle, Einstein arrived at the field equations of general relativity (see the box opposite) which relate the curvature of space-time to the mass–energy density. These equations are too complicated for us to write down in detail, but their content may be summarized as follows:

the curvature of space-time is proportional to G times the density of energy and momentum where G is Newton's gravitational constant. The field equations determine the curvature. The equivalence principle then tells how matter responds: bodies in free-fall follow the geodesics – 'straight lines' – of the surface.

In Newtonian gravity, the gravitational field is determined by the distribution of matter in space – the matter density. At any point in space, this is just one value. In Einstein's gravity, things are more complicated. First of all, matter, acted on by forces, is in motion, and this generates a momentum distribution. According to relativity, for moving observers energy and momentum become mixed up, just like space and time. Since mass is energy, and energy and momentum are partners, it should not be too surprising that general relativity requires more than just the matter density at each point. The situation is similar to that in electromagnetism, where magnetic fields are generated by electric currents, i.e. moving charges. In gravity, gravitational fields are generated by momentum currents, i.e. moving matter, although the complete analysis is too complex to describe here. Indeed, Einstein's theory requires ten quantities at each point to describe moving matter in a gravitational field.

Momentum-dependent effects are not present in Newton's theory of gravity, so we must ensure that our new theory of gravity is still able to reproduce all of the successes of its Newtonian predecessor. Apart from the matter density, the other nine measurements needed to describe moving

Einstein's field equations

Ten quantities at each point are needed to specify the energy and momentum distribution. How do we specify the curvature of space-time? The details of the mathematics are complicated. Once more we apologize that we have to say 'it turns out' that there are ten quantities required at each point to specify this curvature – the 'Einstein curvature'. Now comes the key observation. Conservation of energy and momentum puts four restrictions on the ten energy–momentum components, which complicates matters further since energy and momentum can be exchanged with the gravitational field. Einstein noted that the restrictions on these energy–momentum components have exactly the same form as four restrictions on the Einstein curvature components. The restrictions on the curvature arise

from the fact that a small change in the coordinate system produces a change in the metric, but that the curvature must be independent of such coordinate changes. Einstein concluded that the energy–momentum density and the curvature of space-time must be proportional. Einstein chose the simplest possible equations and ensured that they gave the same predictions as Newton's theory for slow-moving matter and small masses.

Einstein's field equations may be written in the form

$$E_{\mu\nu} = \frac{8\pi G}{c^4} T_{\mu\nu}$$

where the subscripts μ and ν stand for the four possible coordinate directions of space and time. The quanti-

ties $E_{\mu\nu}$ and $T_{\mu\nu}$ are called *tensors*. A tensor is a generalization of a vector, and both $E_{\mu\nu}$ and $T_{\mu\nu}$ have $4 \times 4 = 16$ components. In fact, since they are symmetric tensors, only ten of these components are independent. The tensor on the left, $E_{\mu\nu}$, is called the 'Einstein curvature tensor'. It is related to the more general 'Riemann curvature tensor' which quantifies the rate at which one geodesic deviates from a nearby geodesic. The 'energy–momentum tensor', $T_{\mu\nu}$, acts as the source of gravity, and the constants in the proportionality are just the Newtonian gravitational constant G and the velocity of light c.

matter all involve the speed relative to the velocity of light. On this scale, the speeds of planets in the solar system are very small. The Earth's speed is only one-ten-thousandth of the speed of light, and we can safely use Newton's gravity for most practical purposes.

Any good scientific theory must be able to make predictions that can be tested by experiment. Einstein was a good physicist, and was therefore well aware of this. On the advice of one of his early mentors in physics, the Dutch physicist Henrik Antoon Lorentz, in March of 1916 Einstein wrote a much more intelligible and coherent account of general relativity. He concluded this paper with a brief section on three experimental tests of general relativity: the red-shift; the bending of light; and an anomaly in the orbit of the planet Mercury. The only experimental confirmation of his theory at that time came from Mercury. We now know that the red-shift test only depends on the equivalence principle and not on the specific relationship between matter and the curvature contained in Einstein's field equations. In fact, the gravitational red-shift was the last of Einstein's predictions to be verified – in 1960 with the Pound–Rebka experiment.

There is one small footnote to be added to the story of the discovery of the field equations of general relativity. It is perhaps not widely known that the great German mathematician David Hilbert published the same equations at more or less the same time as Einstein. Einstein had visited Göttingen earlier in 1915 and had given six two-hour lectures there. The subject was general relativity, and Einstein wrote:

> To my great joy I succeeded in convincing Hilbert and Klein completely.

This was before Einstein had made his final breakthrough in October and November of that year. In November, Einstein and Hilbert had corresponded by letter about some problems with Einstein's earlier paper. After a last struggle, Einstein was able to present the final version of his gravitational field equations to the Prussian Academy on November 25th. Five days earlier, Hilbert had submitted a paper containing the same field equations to a journal in Göttingen. For a short time there was clearly some friction between the two scientists, and it is probable that Einstein felt that Hilbert had unconsciously plagiarized his ideas. Einstein then wrote to Hilbert promising to put any possible feelings of 'pique' aside. In this way, Einstein and Hilbert were able to surmount this difficult period in their relationship and continue their friendship. A final remark by Hilbert about Einstein summarizes the true situation very well:

> Every boy in the streets of Göttingen understands more about four-dimensional geometry than Einstein. Yet, in spite of that, Einstein did the work and not the mathematicians.

Let us now look at the experimental evidence for Einstein's other two tests – the bending of light and the orbit of Mercury. We go on to consider a fourth test of general relativity not considered by Einstein – the time delay of light.

Figure 9.10 David Hilbert (1862–1943) is recognized as being one of the great mathematicians of all time. He was the inventor of 'Hilbert space' – an infinite-dimensional analogue of Euclidean space – which forms the mathematical basis of quantum mechanics. At the 1900 International Congress of Mathematicians in Paris, he proposed twenty three unsolved problems for mathematics. Hilbert also initiated a programme for placing mathematics on an incontestably sound basis. In the 1928 Mathematical Congress in Bologna, Hilbert distilled his programme into three fundamental questions: Was mathematics complete? Was mathematics consistent? Was mathematics decidable? In 1930, Kurt Godel demolished Hilbert's programme by showing that arithmetic must be incomplete: there must exist assertions that could neither be proved or disproved. Another great mathematician, Von Neumann, was giving lectures on Hilbert's programme when he read Godel's paper: he cancelled the rest of the course and instead gave lectures on Godel's work. The rout of Hilbert's programme was completed by Alan Turing, who showed with a simple mechanistic model that there were genuinely unsolvable problems in mathematics. Hilbert is said to have remarked: 'Physics is much too hard for physicists'. His tomb is engraved with the epithet: 'We must know. We will know'. [Deutsches Museum, München.]

We begin with the effect that gave Einstein world-wide fame – the bending of light.

Mirages in space

Dear Mother, joyous news today. H. A. Lorentz telegraphed that the English expeditions have actually demonstrated the deflection of starlight from the Sun.

Albert Einstein, postcard to his mother, September 27th, 1919

In chapter 8, we saw how Einstein realized that his equivalence principle would lead to the bending of light. His first prediction concerning the deflection of starlight passing close to the Sun was given in 1911 – before he had completed the theory of general relativity. The magnitude of the effect (less than 1 arcsecond, where 1 degree contains 3600 arcseconds) proved to be equal to that obtained by applying Newtonian gravity to light, assuming light to be a stream of particles. Unknown to Einstein, this result had been obtained over one hundred years earlier by the German astronomer Johann Georg van Soldner. The fact was used by Nazi physicists, including two Nobel prize winners Philipp Lenard and Johannes Stark, in their attempts to discredit 'Jewish science'. Yet, in November 1915, Einstein revised his prediction. Using his new theory of general relativity, he had found that the correct deflection was exactly twice his previous result.

This doubling comes about because, in Einstein's theory, space itself, as well as space-time, is curved. The French mathematician Elie Cartan showed that Newtonian gravity was equivalent to a *flat* space but a *curved* space-time. We can see the reason for this extra contribution from Einstein's theory in a number of ways. One way to explain the effect is to note that Einstein's general relativity predicts that there should be an additional contribution due to the motion itself; i.e. that the momentum, as well as the energy, must be taken into account. For light, the energy and momentum are equal (using appropriate units), and this argument correctly suggests that general relativity should predict twice the Newtonian value for the deflection. Another, more intuitive, way to understand the extra deflection attempts to visualize curved space near the Sun. A two-dimensional section of curved space in the vicinity of the Sun can be pictured as an elastic surface with a heavy ball sitting on it to represent the Sun (see Figure 9.11). Light rays follow the 'geodesics', shortest paths, of this surface. Einstein's first calculation gave the predicted deflection of light relative to locally straight lines. From such a picture, it is possible to convince oneself that the bending of these 'straight' lines due to the curvature of the space around the Sun will increase the Newtonian answer. The critical question is the amount by which the result will be increased. The Einstein field equations of general relativity reveal the answer: the result is an additional deflection by an amount exactly equal to his earlier prediction. This prediction tests Einstein's

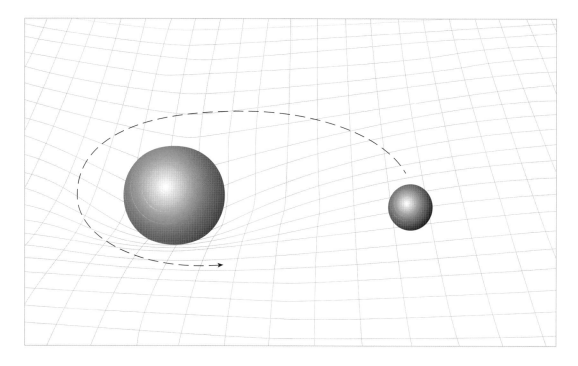

Figure 9.11 Arthur Eddington pictured the distortion of space by a massive body in terms of a heavy ball deforming a rubber sheet on which it is placed. A smaller ball rolling on this surface will respond to the curvature of the sheet as if there were an attractive force between the two objects.

theory because other possible theories of gravity predict a different curvature and a correspondingly different additional deflection. The 'Brans–Dicke' theory of gravity, which was proposed as an alternative to Einstein's, gives a different prediction for the bending of light. As we will see later, the Brans–Dicke theory, although it challenged Einstein's theory during the 1960s and 70s, is now no longer considered a serious alternative to general relativity.

How can so small a bending of starlight be measured? A ray of light passing near the edge of the Sun can only be observed during a total eclipse. An attempt by the German astronomer Erwin Finley-Freundlich to mount an expedition to test Einstein's earlier (incorrect) prediction had to be abandoned when the First World War was declared. Europe was at war by the time Einstein had arrived at his general theory, in November of 1915, with the revised prediction for the deflection. Although there was no direct communication between German and British scientists during the war, the Dutch physicist Willem de Sitter forwarded Einstein's paper to Arthur Eddington in

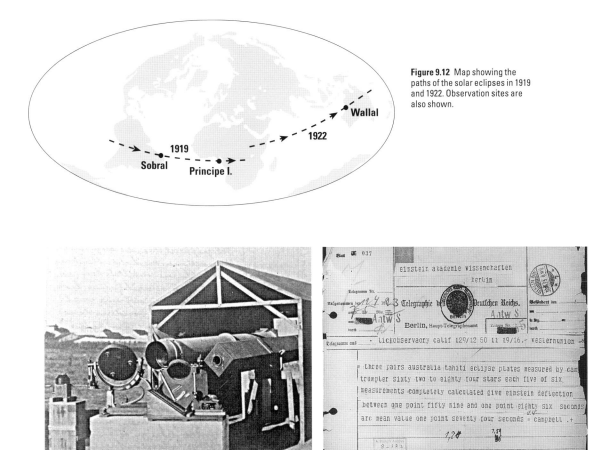

Figure 9.12 Map showing the paths of the solar eclipses in 1919 and 1922. Observation sites are also shown.

Figure 9.13 Optical instruments used to study the eclipse at Sobral in northern Brazil.

Figure 9.14 Telegram sent to Einstein to confirm the results of the 1922 eclipse measurements. [Permission granted by The Albert Einstein Archives, The Jewish National and University Library, The Hebrew University of Jerusalem, Israel.]

Cambridge. Eddington immediately recognized the power and importance of Einstein's theory, and he began planning an expedition to measure the predicted deflection of starlight. The best opportunity was given by an eclipse of May 29th, 1919, during which the Sun would be passing in front of the large number of bright stars in the Hyades cluster. The outcome of the war was still in doubt when permission and support were requested from the British government. Eddington was a devout Quaker who had been granted exemption from military service as a conscientious objector. The Ministry of National Service tried to overturn this exemption but, after three hearings and special pleading from the Astronomer Royal, Frank Dyson, Eddington was allowed to go on the expedition.

Total eclipses have a habit of being observable in very remote places, and this was no exception. Arthur Eddington led one party to the island of Principe off the coast of West Africa; a second party under Andrew Crommelin was sent to Sobral in northern Brazil. Another problem with eclipses and Earth-bound astronomy is that the best-laid plans can be thwarted by the weather. On the day of the eclipse, Eddington was very anxious when 'a tremendous rainstorm came on'. Fortunately, during the brief five-minute period of total eclipse, the sky cleared and Eddington was able to take some photographs. Only two plates had reliable star images. The measured star positions had then to be compared with the positions of the stars on a photograph of the same area of sky taken at another time of year when the Sun was elsewhere. Eddington had with him the results of such measurements, and, when he compared the two sets of measurements, he found that the deflection was in agreement with Einstein's predictions. He later said: 'That was the happiest moment of my life!'

The expedition to Sobral was more fortunate, and Eddington was able to conclude:

> The evidence of the Principe plates is just about sufficient to rule out the possibility of the 'half-deflection' [Einstein's 1911 result without the curved space contribution], and the Sobral plates exclude it with practical certainty.

The results of the eclipse expedition were presented to a packed meeting of the Royal Astronomical Society on November 6th, 1919. At the meeting, Ludwick Silberstein stood up and pointed to a portrait of Newton on the wall, saying:

> We owe it to that great man to proceed very carefully in modifying or retouching his Law of Gravitation.

This did not prevent the press enthusiastically taking up Einstein as a symbol of a new post-war age. The *New York Times* reported

> Lights all askew in the heavens: Men of science more or less agog over results of eclipse observation.

Figure 9.15 The cover of the magazine *Berliner Illustrirte Zeitung* published December 14th, 1919. The caption reads: 'A new great figure in world history: Albert Einstein, whose investigations signify a complete revision of our concepts of Nature, and are on a par with the insights of a Copernicus, a Kepler, and a Newton.'

Unusually for a scientist suddenly thrust into the spotlight of world-wide media attention, Einstein was able to live up to this iconoclastic image. His research career had started in obscurity, outside the scientific establishment, and he was now challenging the established truths of Newton and Euclid. Couple this with Einstein's cheerful manner and his willingness to talk to the press on any subject, ranging from science to religion and politics, and it is easy to see some of the reasons why he became such a public figure. If he had lived in our present day and age, it is clear that he would have provided memorable 'soundbites'. For example, Einstein used to be seen incessantly cleaning his pipe, and, when someone asked him whether he smoked for the pleasure of smoking or in order to engage in the process of unclogging and refilling the pipe, he replied:

> My aim lies in smoking but as a result things tends to get clogged up I'm afraid. Life, too, is like smoking, especially marriage.

Although Eddington's analysis of his eclipse measurements supported Einstein's prediction, it was limited to 10 to 20% precision. Until 1973, there were numerous eclipse expeditions, but, in spite of considerable advances in technology, only a modest increase in accuracy was obtained. By 1973, such optical measurements had been superseded by another, much more accurate, technique.

Like light, radio waves are part of the spectrum of electromagnetic waves predicted by Maxwell's equations. General relativity predicts that radio waves will be deflected by a gravitational mass in just the same way as light waves. In retrospect, it seems surprising that physicists and astronomers paid little attention to the possibilities of radio astronomy until 1931, when Karl Jansky of the Bell Telephone Laboratories first identified radio noise coming from the centre of our galaxy. Radio noise from the Sun was not discovered until 1942, when the young British scientist, James Stanley Hey (no relation to one of the authors!), realized that the jamming of the British early warning radar system was due not to the Nazis but rather to interference from solar radio emissions. This discovery was immediately classified as a military secret for the remainder of the war. Perhaps the reason for the lack of interest in radio astronomy was due to the relationship between wavelength and the resolving power of a telescope. To understand this point, consider what we see when we look at an oil painting from different distances. From a distance, we are unaware of the individual brush strokes and detailed texture of the canvas. As we approach the picture more

Figure 9.16 Two of the radio telescopes of Caltech's Owens Valley Observatory that were used in the pioneering measurements of the deflection of quasar radio waves by the Sun. The two telescopes are separated by just over 1 kilometre, and the dish of the foreground instrument is 130 feet in diameter, while the background antenna has a diameter of 90 feet. [Caltech's Owens Valley Radio Observatory.]

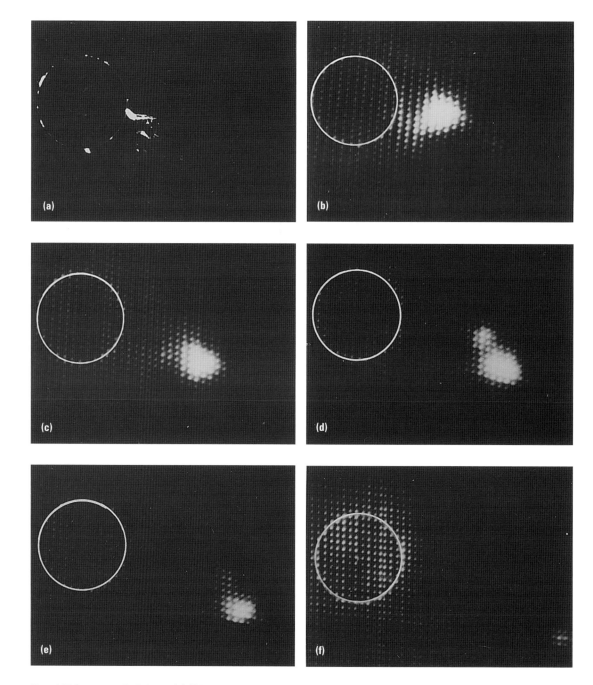

Figure 9.17 A sequence of radio images (b)–(f) following the eruption of a flare prominence shown in the optical image (a). These pictures show a cloud of plasma ejected from the Sun and travelling at a few hundred kilometres per second. Terrestrial magnetic storms and the aurora may follow a day or two after the occurrence of a large flare when the pulse of plasma arrives in the vicinity of the Earth. [CSIRO Radiophysics Division.]

Figure 9.18 A radar image from the European Space Agency's ERS-1 satellite, of waves moving through the Straits of Gibraltar. The wavelength is about 2 kilometres. The spreading of waves emerging on the right into the Mediterranean is caused by diffraction from the narrow gap of the Straits. All waves behave in this fashion, and it is this fact that limits the resolution of telescopes and microscopes. [Copyright ESA ERS (1992).]

closely, we begin to be able to make out these features. We say that we are now able to 'resolve' details at the scale of brush strokes. The same result could also be obtained by standing still and looking at the picture from a distance through a telescope. In the case of a work of art this would be absurd: for astronomical objects, this is all we can do. The resolving power of a telescope is therefore crucial for determining the level of detail we are able to observe. Under ideal conditions, this depends on two things: the size of the primary lens or mirror, and the wavelength of the radiation from the object we are trying to observe. Doubling the size of the mirror doubles the resolving power of the telescope. Doubling the wavelength has the opposite effect and halves the resolution. For this reason, since visible light has a much shorter wavelength than radio waves, radio telescopes were expected to be poor at resolving detail and of little use for accurate measurements. How is it then that radio astronomy has contributed some of the finest resolution measurements in modern astronomy?

The answer lies in the use of the 'radio interferometer'. This consists of two radio telescopes separated by a distance which can be many kilometres long. This distance is called the 'baseline'. The telescopes both receive radio waves from the same source, but, depending on the direction of the source,

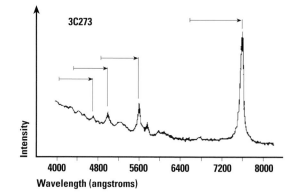

Figure 9.19 The spectrum of quasar 3C 273 showing four bright emission lines due to hydrogen. The arrows indicate the red-shift: the magnitude of the Doppler shift indicates that the quasar is moving away from us at about 15% of the speed of light. (10 000 angstroms are equal to one-millionth of a metre.)

the signal has to travel further to one of the telescopes. When the signals are recombined, the difference in path lengths shows up as a typical wave interference effect, in similar fashion to the famous Michelson and Morley experiment for light. Since only a tiny change in direction can alter a cancellation of two waves into a reinforcement, such radio interferometers can have very large resolving power. The 'Goldstack' interferometer, combining the telescope at Goldstone, California, with the Haystack Telescope in Massachusetts has a baseline of around 3900 kilometres. 'Intercontinental' interferometers using several radio telescopes around the world have also been used. Resolutions of less than 1/1000 arcsecond have been achieved in this way. This is equivalent to distinguishing between two fireflies 25 millimetres apart at a distance of 8000 kilometres! Since Einstein's predicted deflection is greater than 1 arcsecond, we would like to be able to use such radio interferometry to measure the deflection to high accuracy. The problem is that we need a sharp source of radio waves – not the radio waves from a galaxy, which can extend over as much as 1 degree. Fortunately, the discovery of quasars ('quasi-stellar objects') in the 1960s has provided us with just the sort of objects that are needed to make the test possible: they are radio-emitters, star-like, and have a small angular size.

For several days at the beginning of October in 1969, radio astronomers at Owens Valley and Goldstone in California waited for the two bright quasars '3C 273' and '3C 279' to pass close to the Sun. On October 8th, quasar 3C 279 passed behind the Sun, and 3C 273 was only about 4 degrees away. The two groups of radio astronomers used the angle between the two quasars to measure the gravitational bending of the radio waves. In 1969, both groups obtained a result in agreement with general relativity, to within about 10% accuracy. This was not much better than the eclipse measurements but, unlike a solar eclipse, this experiment could be repeated each October. By 1975, using quasars 3C 273 and 3C 279, and also another trio of quasars, radio astronomers had improved the accuracy of this test of general relativity to 1%.

Figure 9.20 Johannes Kepler (1571–1630) was one of the first advocates of Copernicus's Sun-centred astronomy. His three laws of motion for the planets are a landmark in the history of science and laid the groundwork for Newton's theory of gravity. Kepler wrote one of the earliest science fiction books, *Somnium*, about a dream journey to the Moon. His mother was also prosecuted for being a witch, and Kepler defended her at her trial. Kepler started his career with the publication of an astrological calender: he ended it as astrologer to General Wallenstein. [Photo AKG London/Erich Lessing.]

The search for Vulcan

The final release from misery has been obtained. What pleased me most is the agreement with the perihelion motion of Mercury.
Albert Einstein, letter, 1915

The greatest success of Newton's theory of gravity was its explanation of the motion of the planets. Johannes Kepler had laid the foundations of 'celestial mechanics' in two books: *Astronomia Nova*, published in 1609, and *Harmonice Mundi*, published in 1619. In these books, Kepler not only adopted the radical and heretical view of Copernicus that the Earth went round the Sun, but went further and demolished the idea of the planets moving in 'uniform motion in perfect circles'. It was the accuracy of the Danish astronomer Tycho Brahe's planetary observations that had finally forced Kepler to abandon Aristotle's perfect circles. Kepler was concerned about a discrepancy of 8 arcminutes in the orbit of Mars:

> …if I had believed that we could ignore these eight minutes, I would have patched up my hypothesis accordingly. But since it was not permissible to ignore them, those eight minutes point the road to a complete reformulation of astronomy: they have become the building material for a large part of this work.

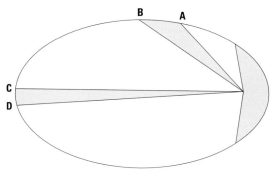

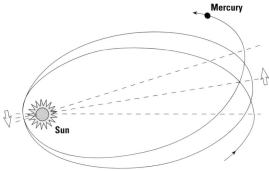

Figure 9.21 Kepler's second law states that a line from the Sun to any planet sweeps out equal areas in equal times. In the figure, the planet takes the same time to travel from A to B as it does to travel from C to D.

Figure 9.22 The slow rotation of the elliptical orbit of a planet, in this case Mercury. The perihelion is the point of closest approach to the Sun, and the gradual shift of this point is called the perihelion precession. (Not to scale.)

Unlike Ptolemy and Copernicus, Kepler did not have the convenience of inaccurate data that he could fit to his prejudices. After an immense struggle, Kepler eventually arrived at the conclusion that planetary orbits were ellipses, with the Sun not at the 'centre' but rather at one focus of this ellipse. Unlike Copernicus and Newton, Kepler wrote down in detail all the twists and turns of the thought processes he took before arriving at this result. He concluded:

> I thought and searched, until I nearly went mad, for a reason, why the planet preferred an elliptical orbit [to mine].... Ah, what a foolish bird I have been!

In his two works, Kepler cleared away an immense amount of historical baggage that had blocked the road to progress in astronomy for over 1000 years. With his three laws he had set the scene for Newton to invent the field of celestial mechanics. Kepler's first law said that planets move in ellipses. His second and third laws are a fundamental landmark for modern physics in that they were precise, verifiable statements expressed in mathematical terms. The second law states that the line joining the Sun and a planet sweeps through equal areas in equal times; the third says that the square of the period of the orbit of a planet is proportional to the cube of its semi-major axis. This last law, with its relationship between the period of the planet and its distance, gave Newton the final clue for his theory of gravity. In his writings, Kepler repeats many times the phrase 'there is a force in the Sun that moves the planets'. It was left to Newton to give a concrete realization to this force. We now come to the story of a much smaller discrepancy than Kepler's 8 arcminutes in the orbit of Mars: Einstein was concerned with 43 arcseconds and the orbit of Mercury.

One of the greatest triumphs of Newton's theory of gravity was the successful prediction of the planet Neptune. John Couch Adams, in England, and Urbain Jean Joseph Le Verrier, in France, independently analysed the orbit of the planet Uranus. To do this, one has to take into account not only the gravitational attraction of the Sun, but also the effect of all the other planets on the motion of Uranus. When all these planetary gravitational forces were included, there was still a discrepancy in the planet's orbit. Both Adams and Le Verrier suggested that this was due to the existence of an undiscovered planet beyond Uranus. On the nights of September 23rd and 24th in 1846, the planet Neptune was discovered at the Berlin Observatory. In 1859, Le Verrier, then director of the Paris Observatory, attempted to repeat this success. His detailed analysis of the motion of Mercury revealed a discrepancy. If one only takes into account the gravitational attraction due to the Sun, the orbit of Mercury would be fixed in space. One of the effects of including the small 'perturbations' caused by the other planets is that the orientation of the orbit is no longer fixed but slowly rotates. This is shown in Figure 9.22. The perihelion of Mercury is the point in the orbit that is closest to the Sun. With these perturbations, the line drawn from the perihelion to the Sun rotates – 'precesses' – so that the orbit traces out a rosette pattern. Mercury's orbit rotates 574 arcseconds per century, and Le Verrier calculated the contributions to this precession from all the planets. The closest planet, Venus, makes the largest contribution of 277 arcseconds. Jupiter, being 400 times more massive than Venus, gives the next largest contribution of 153 arcseconds. Earth contributes another 90 arcseconds, and Mars and the rest of the planets amount to a further 10 arcseconds. This leaves 43 arcseconds which cannot be accounted for by this analysis. Anticipating a similar success to that concerning the discovery of Neptune, a new planet, Vulcan, was proposed to explain the discrepancy. This was predicted to be closer to the Sun than Mercury and hence was named after Vulcan, the Roman god of fire, whose duties included forging thunderbolts for Jupiter. A search was made for Vulcan, but this was without success. The precession of the perihelion of Mercury remained an outstanding problem for astronomy until Einstein used his new theory of gravity to explain the missing 43 arcseconds.

In general relativity, one can identify three different effects, not in Newtonian gravity, that contribute to the perihelion precession of Mercury. First there is the velocity dependence of the gravitational force. As a planet moves towards the Sun, it speeds up. The velocity-dependent force in Einstein's gravity therefore causes the planet to go a little further round the Sun than would be the case for purely Newtonian gravity. The second effect is the curvature of space. The curved-space distance around the Sun is different from its flat-space value, so that the trajectory does not quite join up in the Newtonian manner. The third effect is the fact that gravity acts on all energy – including the energy of the gravitational field itself. This is characteristic of a 'non-linear' theory. Although we have presented these three effects as distinct, in reality they are not so clearly separable. In spite of this ambiguity, the final result of the shift predicted by general relativity is unambiguous: almost exactly 43 arcseconds per century. When Einstein realized that this result was a natural consequence of his theory, 'without any special hypothesis', he was overjoyed. Abraham Pais writes that:

> This discovery was, I believe, by far the strongest emotional experience in Einstein's scientific life, perhaps in all his life. Nature had spoken to him.

In the 1960s, the use of modern computers and new radar data verified the perihelion predictions of general relativity, not only for Mercury but also for Venus, Earth and the asteroid Icarus. The 1960s also brought another development that cast some doubt on this classic test of general relativity. This challenge to Einstein began in 1960 when Robert Dicke and his graduate student at Princeton, Carl Brans, put forward an alternative theory of gravity. This theory incorporated the same equivalence principle as Einstein's theory, but differed in its prediction for the curvature of space-time. As a result, the 'Brans–Dicke' theory, as it was called, predicted a slightly smaller precession of Mercury than that predicted by Einstein. In order to try to account for this difference, Dicke and H. Mark Goldenberg began a series of measurements to determine if the Sun is exactly spherical. If the Sun is slightly flattened at the poles, this 'oblateness' will cause a further contribution to the perihelion precession of a planetary orbit. When Dicke and Goldenberg first reported their results in 1966, they concluded that the solar

Figure 9.23 One of the many complex oscillation patterns for the Sun: red areas are receding and blue areas are approaching. [NOAO.]

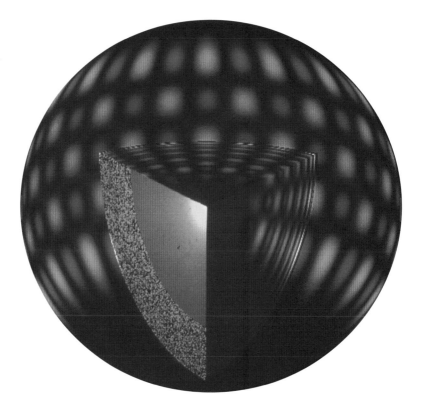

flattening was large enough to contribute about 3 arcseconds each century to the perihelion shift of Mercury. Their results sparked a storm of controversy that has still not been fully resolved, although most astronomers now believe that the solar flattening is much less than the original amount proposed by Dicke and Goldenberg.

One remarkable result that emerged from this renewal of interest in the solar surface was the discovery of solar vibrations. In 1976, Henry Hill and his colleagues, working at the Santa Catalina solar observatory near Tucson in Arizona, discovered that the Sun is in continual vibration. The solar surface undergoes regular distortions, with periods ranging from several minutes to several hours. These new uncertainties in the physics of the Sun complicate the picture further. Brans–Dicke theory has come and gone, but enough questions have been raised to make the perihelion test of general relativity not quite the rock-solid confirmation of the theory it once seemed to be.

General relativity and the velocity of light

In 1964, a fourth test of general relativity was proposed by the US physicist Irwin Shapiro. He was working at MIT on improved measurements of planetary distances using the new technique of 'radar ranging'. This involved measuring the time interval between the emission and reception of radar signals bounced off the surface of a planet. In 1961, this technique had been used to measure the distance of the closest approach of Venus to the Earth. The returning echo was very weak, but detectable, and, when combined with the speed of light, the time taken gave a measurement of the distance to within an accuracy of about 100 kilometres. Prior to these radar measurements, the uncertainties in interplanetary distances were in the range of tens of thousands of kilometres. Shapiro had the idea of using radar ranging to test general relativity. If the path of the radar signal passed close to the Sun, there would be a delay due to the gravitational time dilation and the equivalence principle. Just as in the case of the perihelion of Mercury, there would be an additional contribution due to the curvature of space. Any time delay caused by the bending of the light path turned out to be negligible.

How big is the predicted effect, and can it be measured? The effect is expected to be largest when the Earth and another planet are in a straight line on opposite sides of the Sun – a situation known as 'superior conjunction'. Calculations for Mars yield a predicted delay for the round-trip time of some 250 millionths of a second or 250 microseconds. Since light travels 75 kilometres in 250 microseconds, this corresponds to a difference in the Mars–Earth distance of half this distance, nearly 40 kilometres. If we can measure the time delay to this accuracy, we are seemingly faced with a choice. Either Mars has moved to an orbit 40 kilometres further away at superior conjunction, and then resumes its original orbit as the planet continues on its way, or we must conclude from the time delay that the orbit is the same but that light travels more slowly in this situation. How do we reconcile

Figure 9.24 Irwin Shapiro received his Ph.D. from Harvard in 1955, and went on to work at nearby MIT on experiments on radar ranging to planets. Attending a lecture in 1961, he was puzzled by the statement that, according to general relativity, the speed of light is *not* constant. After looking up the equations of general relativity, Shapiro applied them to radar ranging and discovered a new, fourth test of general relativity. [Courtesy of the Harvard-Smithsonian Center for Astrophysics.]

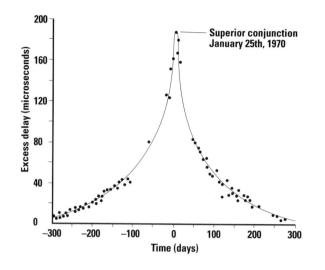

Figure 9.25 A graph of the excess time delay of radar echoes from Venus. The maximum time delay – labelled 'Superior conjunction' – occurs when Venus, the Sun and the Earth are all in line, with Venus on the far side of the Sun, so radar travelling to and fro passes closest to the Sun.

Figure 9.26 A mock-up of the Mariner 6 spaceprobe, which was the reflector in a 1970 measurement of the Shapiro effect on radar time delay. [Provided through the courtesy of the Jet Propulsion Laboratory, California Institute of Technology, Pasadena, California.]

Figure 9.27 A view of the Utopia plains of Mars taken from the Viking Lander 2. [Provided through the courtesy of the Jet Propulsion Laboratory, California Institute of Technology, Pasadena, California.]

this apparent slowing down with the fact that the velocity of light is supposed to be the same in every small free-falling region? In fact, there is no problem because we must compare a whole sequence of such regions along the light path to determine the time taken. All that the observer can say is that the light takes longer to complete a round-trip when the light path goes near the Sun. We can describe this as light slowing if we wish, but the only *measurable* quantity is the time *delay*. Shapiro had first calculated his time delay predictions in 1961 and 1962, but had concluded that current radar antenna were not powerful enough to generate a measurable echo over the large distances required.

The situation changed in 1964 with the completion of the Haystack radar installation: a 250 microsecond delay was, in principle, detectable, provided that radar ranging could be made accurate to a few tens of microseconds. After Shapiro published his paper predicting the delay, the power of the Haystack radar was increased by a further factor of 5. In 1966 and 1967, measurements were made for both Mercury and Venus, and the results supported general relativity to within an uncertainty of about 20%. By the end of 1970, the accuracy of the measurements had been improved, and they now agreed with Shapiro's prediction to within 5%. It is difficult to do much better with planetary radar because of uncertainties associated with mountains and valleys on the surfaces of the planets.

At the Jet Propulsion Laboratory in California, scientists realized it should be possible to measure the time delay using spaceprobes instead of planets. After the *Mariner 6* and 7 missions to Mars had completed their studies, several hundred range measurements were made to each of the spacecraft at and near their superior conjunctions in 1970. These time delay measurements agreed with the predictions of general relativity to within 3%. The most accurate measurement of the time delay came from combining the planet and spacecraft techniques. In 1976 came the *Viking* mission to Mars: two spacecraft landed on the surface of Mars and sent back spectacular

A brief history of radio astronomy

In the closing years of the nineteenth century and the beginning years of the twentieth century, radio waves were still a novelty, and new experiments and applications were pursued with enthusiasm. Several scientists tried to detect radio waves from space and from the Sun. They were unsuccessful for two reasons. First, the long-wavelength radiation they attempted to detect is reflected from the conducting layers of the upper atmosphere and does not reach the ground. Secondly, although the Sun does generate radio waves, it, along with most other stars, is a relatively weak source of such radiation. Thus, it was not until the Second World War that radio waves from the Sun were first detected, and then only by a war-time 'accident'. Between February 26th and 28th, 1942, the British early warning radar system became 'jammed'. At first, this was attributed to hostile action on the part of the Germans. It was James Hey, a young civilian scientist working in the war-time radar research programme, who realized that this interference was, in fact, coming from the Sun. In addition, he correctly proposed that this unusual radio noise from the Sun was associated with a large solar flare which had occurred at the same time. News of this discovery was restricted during the war for reasons of military security.

Cosmic radio signals had been discovered somewhat earlier by a radio engineer named Karl Jansky. Jansky was working for the Bell Telephone Laboratories in New Jersey, and was given the task of tracking down the cause of interference on trans-Atlantic radio telephone reception. In late 1931, he built a steerable radio antenna tuned to relatively short radio wavelengths. Three sources of radio noise became apparent: local thunderstorms, more distant thun-

Karl Guthe Jansky (1905–1950) with the antenna with which he demonstrated that the Milky Way was a bright radio source. His discovery was made in 1931, but received very little attention from astronomers. [Courtesy NRAO/AUI.]

derstorms and a steady hiss that varied with the time of day. At first, Jansky thought the third source must be the Sun, but he eventually realized that the greatest signal always occurred when his aerial pointed in a direction fixed relative to the stars. This direction turned out to be the direction of the centre of our galaxy: Jansky was listening to signals 'broadcast' from the Milky Way. Surprisingly, Jansky's discovery was ignored by all the professional astronomers, and it was left to another radio engineer, Grote Reber, to take the next steps.

In 1937, Reber built a steerable, 9 metre, bowl-shaped radio reflector with an antenna at its focus. This was the world's first radio telescope, and it was located in the backyard of Reber's house in Wheaton, Illinois. Reber made the first radio maps of the sky, and, in addition to the peak of activity near the galactic centre in Sagittarius, he discovered smaller

Grote Reber standing in front of the radio telescope he built in 1937. The world's first radio telescope was built in the back yard of his house in Wheaton, Illinois. [Courtesy NRAO/AUI.]

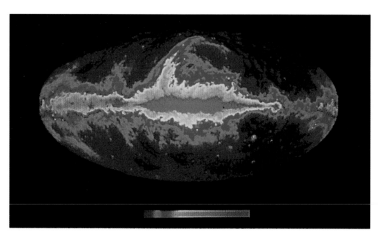

Colour-coded map of the radio sky at a wavelength of 73 centimetres with red corresponding to the brightest regions and black to regions where the radio sky is almost completely dark. The plane of the Milky Way lies across the central band, and shows that the central Sagittarius region in the direction of the centre of the galaxy is the most intense source of radio waves.
[G. Haslam *et al.*, Max-Planck-Institut fuer Radioastronomie (Germany), using observations from Effelsberg, Jodrell Bank and Parks; colour radio map produced at Computing Centre of Rheinisches Landesmuseum, Bonn, by R. Beck.]

James Stanley Hey was on the staff of the UK Army Operations Research Group from 1940 to 1952. He then became a research scientist at the Royal Radar Establishment in Malvern, and was eventually promoted to chief scientific officer. In addition to his discovery of radio noise from the Sun, in 1946 Hey and his colleagues identified the first known extra-galactic radio source, Cygnus A.
[Crown copyright is reproduced with the permission of the Controller of HMSO.]

peaks of radio emissions in Cassiopeia, Cygnus and Taurus. The radio source in Cassiopeia was subsequently associated with the remains of a star that had exploded hundreds of years earlier. The source in the constellation of Taurus, centred on the Crab nebula, turned out to be of a similar nature. This was in the position at which Chinese astronomers observed a sudden flare-up of a new star in 1054 AD. The 'supernova', as we now call such events, was so bright as to be visible in daylight for several weeks. Supernova remnants are now known to be among the brightest objects in the radio sky. As we shall see, the source in Cygnus turned out to be even more interesting. Reber's work laid the foundations for the new field of radio astronomy, and, until the end of the war, Reber was the world's only radio astronomer.

After the war, some of the pioneers of radar moved into radio astronomy. Bernard Lovell set up the Manchester University centre at Jodrell Bank, and Martin Ryle established the Cambridge University radio astronomy group. In 1946, James Hey and his colleagues used old army radar equipment to discover the first extra-galactic radio source, Cygnus A. This was the result of a deliberate attempt to improve Reber's radio map. The radio noise from the Cygnus region was found to vary rapidly in a way quite unlike the emission from the rest of the Milky Way. By 1954, a more precise position for Cygnus A was obtained. Walter Baade and Rudolf Minkowski used the 200 inch Mount Palomar optical telescope to identify Cygnus A optically. What they saw was an object more than 1000 million light years away that appeared to be

two galaxies in collision. Cygnus A was the first of a new class of very powerful radio transmitters now called 'radio galaxies'.

During the 1950s and 1960s, one of the major tasks of radio astronomy was to make lists of the positions of radio sources in the sky. The radio astronomers in Cambridge compiled several such lists, and in 1959 they brought out their third list – the *3rd Cambridge Catalogue* – containing 471 sources. Some of the entries corresponded to objects in our own galaxy, such as the source associated with the Crab nebula. In 1960, Allan Sandage discovered a 'star' at the location of source '3C 48' – i.e. the 48th entry in the *3rd Cambridge Catalogue*. This was peculiar, since it was known that ordinary stars produce little radio noise. In addition, the

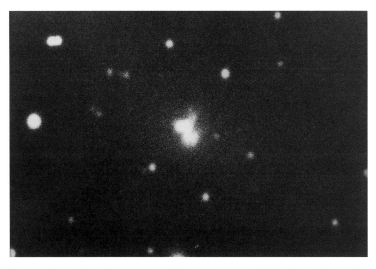

An optical image of the radio galaxy Cygnus A. The radio energy emitted from Cygnus A is roughly equal to its optical emission, which is about one million times the radio energy produced by the Andromeda Galaxy. Moreover, the radio energy from Cygnus A is concentrated in two giant lobes separated by about half a million light years on either side of the central galaxy. Radio emission from a normal galaxy like Andromeda is mostly concentrated within the area of the optical image of the galaxy. [Palomar Observatory, Caltech.]

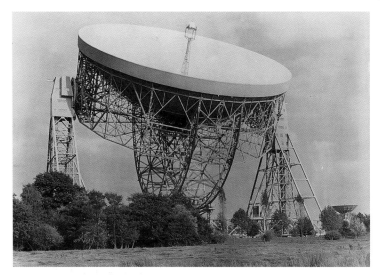

The Jodrell Bank 250 foot steerable radio telescope was completed in 1957, shortly before Sputnik 1 – the first artificial satellite – was launched by the Russians. When equipped with a radar transmitter, Jodrell Bank was the only instrument capable of detecting the faint reflected signal from the carrier rocket: radio signals from Sputnik itself were easy to pick up. Jodrell Bank and its director, Bernard Lovell, became famous. [The University of Manchester.]

spectrum was very strange and included a number of spectral lines that no one was able to identify.

Meanwhile, at the Australian National Observatory, Cyril Hazard and several of his colleagues used the Moon to help them obtain accurate positions for radio galaxies. The Moon obscures a disc of the sky, and, as it moves, new stars are eclipsed and others re-emerge. This is called lunar occultation. By careful timing of the decrease of radio brightness from radio galaxies as lunar occultation begins, the position of the source can be accurately determined. In 1962, three occultations of 3C 273 identified its position to such accuracy that an optical counterpart could be located. Hazard discovered that the source had two components. Optical photographs show a 'star-like' object from which a jet is emerging. The spectrum of 3C 273 was puzzling and had many of the mysterious features associated with 3C 48.

The puzzle was resolved in 1963 by the astronomer Maarten Schmidt. He made the dramatic discovery that the spectral lines of 3C 273 were 'normal' apart from an unexpectedly large redshift. From the magnitude of this Doppler effect, he calculated that 3C 273 must be moving away from us at about 15% of the speed of light. (See Figure 9.19.) It was quickly realized that the spectrum of 3C 48 could be understood in the same way: it is receding from us with a velocity about 33% of light speed. If these objects are part of the ordinary expansion of the universe, they must be very distant and extremely powerful radio sources. Because of their concentrated star-like character they were named 'quasi-stellar' radio sources, or quasars.

pictures of the planet's surface. These spacecraft provided an ideal fixed source of radio waves for the time delay experiment. Ranging data were taken for about a year, and the results were analysed carefully. The result, announced in 1978, was in agreement with Shapiro's predictions from general relativity to within 1 part in 1000, or 0.1%.

Postscript

General relativity has now been subjected to far more detailed scrutiny than was provided by Eddington's first eclipse photographs. Exposure to the full force of modern technology has strengthened our belief in Einstein's theory, despite uncertainties raised by measurements of the oblateness of the Sun. The same cannot be said of the rival Brans–Dicke theory. Robert Dicke and Carl Brans proposed modifying Einstein's theory by the addition of an extra field. The relative importance of this new field compared with Einstein's original gravitational field was described by an adjustable parameter of the theory. In the early 1960s, the new theory was consistent with all of the experimental tests of general relativity at their present level of accuracy. The fall of Brans–Dicke theory is a good illustration of the way physics should make progress. The full story is vividly chronicled in Clifford Wills' book *Was Einstein Right?* In outline, the story is as follows.

In 1967, Kenneth Nordtvedt, a university professor working in Bozeman, Montana, discovered a new prediction of the Brans–Dicke theory. Nordtvedt showed that the application of the equivalence principle in Brans–Dicke theory depended on whether the object was laboratory sized, like the small balls of aluminium and platinum used in the Eotvos experiment, or whether it was a massive body the size of the Moon or a planet. For the small objects, the energy of the gravitational attraction is negligible: for the Moon and planets this gravitational energy is significant. In Einstein's theory, the acceleration of both large and small objects turns out to be the same. In Brans–Dicke theory, gravitational energy falls at a slightly different rate than other forms of energy so that the usual equivalence principle breaks down for objects with significant amounts of gravitational self-binding. Nordtvedt found he could test the two theories using accurate measurements of the orbit of the Moon. In 1975, laser-ranging data of the Moon–Earth distance were analysed by two independent groups of scientists. The results showed no evidence for the 'Nordvedt effect' predicted by Brans–Dicke theory. Ironically, one of the physicists involved in the analysis of the data was Robert Dicke.

10 The Big Bang, black holes and unified fields

The sceptic will say 'It may well be true that this system of equations is reasonable from a logical stand-point, but this does not prove it corresponds to Nature.' You are right, dear sceptic. Experience alone can decide on the truth.

Albert Einstein, commenting on 'Unified Field Theory', 1950

The expanding universe

While on Mount Wilson, Einstein and his wife Elsa were given a tour of the observatory. It was explained to them that the giant telescope was used for determining the structure of the universe, to which Elsa replied: 'Well, well! My husband does that on the back of an old envelope.'

From *Man Discovers the Galaxies*, 1930

After his stunning success with general relativity, Einstein began to think about the implications of his theory for the universe as a whole. In 1917, he wrote a paper that began a new field of physics, that would be called 'relativistic cosmology'. He wrote to his friend Paul Ehrenfest:

> I have... again perpetrated something about gravitation theory which somewhat exposes me to the danger of being confined in a madhouse.

Although Einstein had the nerve to provide the first mathematical model of the universe, he had also, in a sense, lost his nerve at the crucial moment. Instead of predicting Hubble's discovery of the expansion of the universe, a solution that followed naturally from his own field equations, Einstein chose to modify gravity and introduce a new repulsive force. This was described by an extra term he added to his gravitational field equations – the so-called

Figure 10.1 Einstein with his first wife, Mileva, and his son Hans Albert in Bern in 1904. [Permission granted by The Albert Einstein Archives, The Jewish National and University Library, The Hebrew University of Jerusalem, Israel.]

'cosmological term'. In a conversation with George Gamow, of whom more later, Einstein said that his introduction of the cosmological term was the 'biggest blunder' he had ever made. In 1917, when Einstein was writing his paper on 'Cosmological considerations on the general theory of relativity', food was in short supply in Berlin because of the war. Einstein was also unwell, and his marriage to Mileva, his first wife, was in difficulties. He was nursed back to health by his cousin Elsa, whom he later married. Despite such problems in his day-to-day existence, it still seems uncharacteristic of Einstein to doubt the predictions of his own theory. Einstein's reluctance to take seriously the radical consequences of his theory may be explained by the status of astronomy in 1917. At this time, it was not only unclear whether there was anything beyond our own galaxy, but also the universe was thought to be a very static place.

Soon after Einstein had published his static solution, Willem de Sitter, the Dutch physicist who had earlier sent Eddington Einstein's paper on general relativity, showed that another, very different solution to Einstein's modified field equations was possible. The next important contribution was made by a remarkable Russian, Alexander Alexandrovitch Friedmann. Friedmann was originally a mathematician but became fascinated by meteorology and the problem of flight. During the First World War, he served with the Russian airforce, helping to develop navigational devices. After the Russian Revolution, Friedmann became a professor and began to

Figure 10.2 The Hooker 100 inch telescope on Mount Wilson. The task of transporting the equipment up the tortuous mountain trail was difficult and hazardous. A truck carrying the telescope tube nearly slid over the edge of a cliff. Milton Humason, one of the great observers at Mount Wilson, was at one time a school 'dropout' who signed up as a mule train driver on the trail supplying the construction at the site. [Courtesy of The Observatories of the Carnegie Institution of Washington.]

study Einstein's field equations. Friedmann showed that if he dropped Einstein's cosmological term he obtained solutions describing expanding matter-filled universes. In 1922, Friedmann sent a paper with his proposals to a German research journal. When the paper was published, it was followed by a critical commentary by Einstein, which began:

> The results obtained in the work cited regarding a non-stationary universe seem suspicious to me ...

Within a year, Einstein had admitted that Friedmann's results were mathematically correct, but he still believed they were not relevant to the static universe observed by astronomers. Friedmann divided his solutions into two classes: one in which the universe would expand forever, and another in which the density of matter was sufficiently large for the gravitational attraction to overcome the expansion and cause an eventual collapse. Tragically, Friedmann died in 1925, at the age of thirty-seven, with the significance of his work unrecognized. The antipathy of many physicists of that time to such attempts at a theory of cosmology is well illustrated by a quotation from the famous English physicist J. J. Thomson. In his memoirs, he wrote:

> We have Einstein's space, de Sitter's space, expanding universes, contracting universes, vibrating universes, mysterious universes. In fact the pure mathematician may create universes just by writing down an equation, and indeed if he is an individualist he can have a universe of his own.

Today, Friedmann's models of the expanding universe form the basis of modern cosmology. In order to understand why, it is necessary to make a brief diversion from our main theme and review the foundations of present-day observational cosmology.

In the early years of this century, building an observatory on top of a mountain was a new idea. In those days, Mount Wilson, overlooking Pasadena and the Los Angeles basin, seemed an ideal site for a telescope, well away from the lights of the city. Nowadays, Los Angeles has grown so enormously that even Mount Wilson is too close to the smog and brightness of the city at night to be of much use for optical astronomy. In 1909, things were different, and a telescope with a 60 inch mirror became operational. This was accompanied eight years later by the famous Hooker telescope – the 100 inch reflector. For many years, this was the largest telescope in the world, and it was with this telescope that Edwin Hubble settled the great debate about the nature of nebulae. The beginnings of this argument about the status of these faint, nebulous objects date back to the mid-eighteenth century. Even as late as the early 1920s, one group of scientists believed that our galaxy, the Milky Way, was all there was to the universe and that nebulae were merely small nearby sub-units of our own galaxy. An opposing group contended that nebulae were actually immense whirlpools of stars well beyond our own Milky Way Galaxy. In 1920, the Swedish astronomer Knut Lundmark had examined the nebulae problem in his doctoral thesis. He concluded that the evidence favoured the extra-galactic interpretation, but that there was still plenty of room for argument. What was needed was a definitive test.

In 1923, Hubble photographed the 'spiral Andromeda nebula' with the 100 inch telescope on Mount Wilson. With the increased resolution provided by the huge telescope he was able to see that the outer regions of the spiral arms were composed of 'swarms of stars'. From the brightness of some of the stars he was able to make some tentative estimates of distance. The really conclusive evidence came from his discovery of so-called 'Cepheid variable' stars in Andromeda and other spiral nebulae. Cepheid variable stars are named after the first observed example of this type of star, Delta Cephei. They are stars whose brightness varies according to a regular cycle, with a period ranging from 1 to 100 days. In 1908, Henrietta Leavitt, working at the Harvard College Observatory, had noticed a relationship between the pulsation period and the luminosity or brightness of the Cepheid stars, which enabled astronomers to estimate their distance. In this way, using Cepheid variables in the Andromeda nebula, Hubble was able to deduce that the nebula was about one million light years away; well beyond the Milky Way Galaxy. The name 'Andromeda Galaxy' has now replaced the old-fashioned 'nebula'. Hubble's original determinations of distances have undergone substantial revisions. In the 1950s, the astronomer Walter Baade showed that the Andromeda Galaxy was significantly further away than Hubble had estimated. Even today, distance determination remains one of the most uncertain areas of cosmology.

The next important step in unravelling the nature of the universe was also taken by Hubble. In 1912, Vesto Slipher of the Lowell Observatory was able to obtain a photograph of the spectrum of the Andromeda Galaxy that was precise enough to allow him to measure a Doppler shift of the spectral lines. He deduced it was approaching us at a speed of about 300 kilometres/second. By 1925, Slipher had accumulated measurements of the Doppler shifts of forty-one galaxies, which showed that most of these galaxies were moving away from us. Then, in 1929, Hubble realized that the speed at which galaxies were moving away from us increased in proportion to their distance. This relation is now known as Hubble's law:

$$v = H \times d$$

where v is the speed of the galaxy and d is its distance. The constant of proportionality, H, is called the Hubble constant.

Using the Hooker telescope on Mount Wilson, Hubble's colleague, Milton Humason, extended the observation of these 'red-shifts' to galaxies moving with speeds of up to 15% of the velocity of light. Hubble's observations had a profound effect on Einstein. In 1930, Einstein travelled to Mount Wilson to meet Hubble, whence he became convinced of Hubble's expanding universe. Einstein formally abandoned his cosmological term in a paper in 1931, referring to the experimental discoveries of Hubble, 'which the general theory of relativity can account for in an unforced way'.

Figure 10.3 (a) The Andromeda Galaxy. (b) The star field in the small white rectangle of part (a). The marked stars are Cepheid variable stars, which were used to set the distance scale of the universe. [Part (a) courtesy UCO/Lick Observatory; part (b) courtesy Palomar Observatory, Caltech.]

Figure 10.4 Henrietta Swan Leavitt (1868–1921) was originally an amateur astronomer. She became a volunteer assistant at Harvard Observatory and joined the permanent staff in 1902. In 1908, while studying stars in the Magellanic Clouds, she noticed that the period of Cepheid variable stars increased with brightness.
[Harvard College Observatory.]

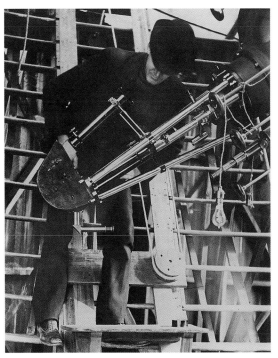

Figure 10.5 American astronomer Vesto Slipher (1879–1969) at the Lowell Observatory in about 1912.
[Lowell Observatory photograph.]

Figure 10.6 (a) Hubble's recession speed versus distance relation was deduced from data on less than twenty isolated galaxies. The black and open circles and the full and dashed line correspond to slightly different ways of treating the same data. (1 parsec = 3.26 light years.) (b) A modern version of Hubble's law data plots the observed red-shift – instead of velocity – against 'apparent magnitude' – related to distance – of distant galaxies. The small rectangle at the bottom left indicates the extent of Hubble's original data. In fact, more than seventy-five years after Hubble's original discovery, although Hubble's law is not in question, the precise magnitude of the so-called Hubble constant – the slope of the straight line – is still hotly debated. [Part (a) is taken from E. Hubble, *The Realm of the Nebula* (Dover Publications Inc., New York). Part (b) is adapted from J. Silk, *The Big Bang* (W. H. Freeman & Co., New York, 1980).]

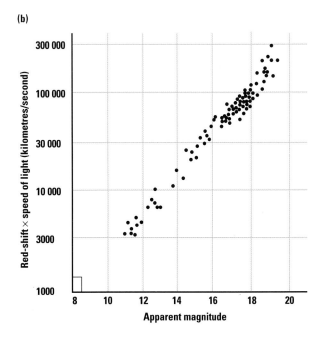

(b)

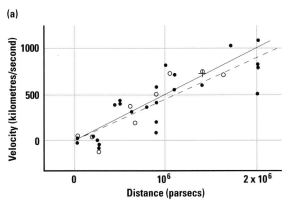

(a)

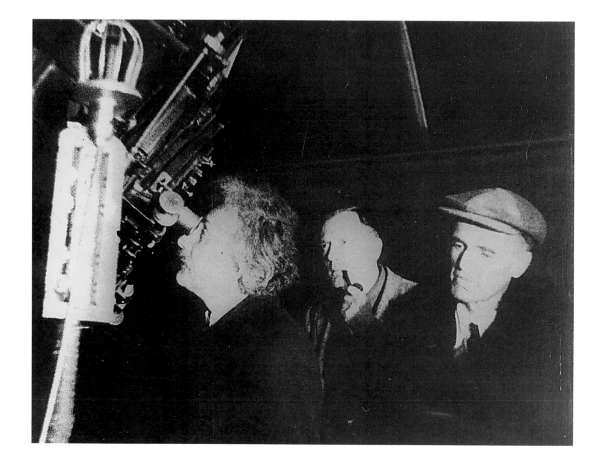

Figure 10.7 Einstein observing at Mount Wilson: Hubble is the man with the pipe. [Reproduced by permission of The Huntington Library, San Marino, California.]

By 1930, the observational work of Hubble had made it clear that the universe was expanding – but Friedmann's solutions to Einstein's unmodified field equations still lay buried and forgotten in the literature. Thus it was that, at a meeting of the Royal Astronomical Society in early 1930, Eddington and de Sitter could agree that no suitable solutions to the field equations had yet been found. Eddington remarked:

> One puzzling question is why there should be only two solutions. I suppose the trouble is that people look for static solutions.

When Georges Lemaître, a Belgian priest and former student of Eddington, read this remark, he wrote to his mentor pointing out that he had published a non-static solution in 1927. Eddington then gave Lemaître's paper high praise and wide publicity, speaking of 'Lemaître's brilliant solution'. Unlike Friedmann, Lemaître was able to relate his solution to the newly discovered expanding universe: the galaxies were flying apart because the fabric of space-time itself was spreading.

Figure 10.8 The Belgian cosmologist George Lemaître (1894–1966) and Einstein in the early 1930s. Lemaître began studying engineering at Louvain University, but his studies were interrupted by the First World War. He served in the army, and after the war he switched his degree course to mathematics and physics. After completing his degree, Lemaître entered a seminary and was ordained as a Roman Catholic priest in 1923. He won a Belgian government scholarship which enabled him to study in Cambridge with Eddington, and he also spent time at MIT and the Harvard College Observatory. He published his solution for an expanding universe in 1927, and in 1931 he was the first to speculate about the explosion of a 'primeval atom' as the origin of this expansion. [Brown Brothers.]

What causes the expansion of the universe? What happens if we run a film of the expanding universe backwards? The Jesuit priest Lemaître was an unlikely candidate to be the first to speculate on the creation of the universe in non-biblical terms. In 1931, Lemaître speculated that the universe may have started from one 'primeval atom' containing the total mass of the universe, and that the observed expansion was caused by the explosion of this cosmic atom, our present universe being the 'ashes and smoke of bright but very rapid fireworks'. This is the 'Big Bang' theory of the universe. Ironically, this catchy name was coined by the British astronomer Fred Hoyle during a BBC radio programme in which he introduced a rival theory – the 'Steady State' model. The implications of the Big Bang theory were first worked out by the Russian-born physicist George Gamow. Gamow had studied under Friedmann, and he considered in detail how nuclear reactions taking place after the primordial fireball of the Big Bang could create the elements we see around us. The key work was carried out by Gamow at George Washington University in the USA with a graduate student named Ralph Alpher. Gamow and Alpher's results were published in the *Physical Review* on April 1st, 1948, with Gamow's friend and later Nobel Prize winner, Hans Bethe, listed as an author. (Including Bethe as an author enabled Gamow to make a pun on the first three letters of the Greek alphabet – alpha, beta and gamma.) Gamow submitted the paper to the journal with Bethe listed as

Figure 10.9 George Gamow (1904–1968), right, and John Cockroft (1897–1967) in about 1930. Gamow was born in Odessa and studied under Friedmann in Moscow. In 1931, after a period of touring research centres in Western Europe, he was appointed as Master of Research at the Academy of Sciences in Leningrad. Two years later, Gamow took the opportunity of attending the Solvay Conference to defect to the West; he eventually settled in America. Gamow is sometimes known as the 'father of Big Bang cosmology', but he also made significant contributions to many other areas of physics, and he also played an important role in the resolution of the genetic code of DNA. Gamow wrote many popular science books, including the famous 'Mr Tompkins' series. [University of Cambridge, Cavendish Laboratory.]

Figure 10.10 Robert Dicke (at right) testing his radiometer, which he used to set an upper limit on the strength of any all-sky radiation, in 1945. Curiously, Gamow's original article on the 'Big Bang' appeared in the same edition of *Physical Review* as the paper by Dicke. In the early 1960s, Dicke and his colleagues at Princeton University realized that there would be microwave background radiation as a relic of the Big Bang. Before they could complete their search for this radiation, Penzias and Wilson, two engineers working for Bell Telephone Laboratories at nearby Holmdel, New Jersey, discovered it accidentally without realizing its significance. Dicke has played a major role in the renaissance of general relativity since the 1960s. With a student, Carl Brans, Dicke proposed a testable alternative to Einstein's version of general relativity. The Brans–Dicke theory stimulated many experiments, but it is now generally agreed that it does not correspond to Nature. [Courtesy R. H. Dicke, Princeton University.]

Figure 10.11 Arno Penzias (right) and Robert Wilson with the horn-shaped radio telescope with which the microwave background radiation was discovered. [Bell Laboratories of Lucent Technologies.]

'*in absentia*'. The editor of *Physical Review* noticed this unusual addition to the author list and sent the paper to Bethe for review. Fortunately, Bethe took the joke in good heart, checked the calculations, removed the words '*in absentia*' from the author list and recommended that the paper be published. In 1948, Gamow's collaborators Ralph Alpher and Robert Herman wrote a paper suggesting that, if the universe started as a fireball, with an estimated temperature of many millions of degrees, there should exist a relic of this fireball in the form of a 'sea' of low-energy electromagnetic radiation. The story of the discovery of this 'microwave background radiation' in 1965 by Arno Penzias and Robert Wilson at the Bell Telephone Laboratories in New Jersey is worthy of a book in itself. At the same time as Penzias and Wilson were performing their experiment, Robert Dicke and James Peebles were independently rediscovering the 'fossil' radiation predicted by the Big Bang. Gamow was understandably upset that in a paper explaining the significance of Penzias and Wilson's 3 degree microwave background radiation, Dicke had not cited Gamow's work nor that of his collaborators, Alpher and Herman. Gamow's subsequent letter to Penzias ends testily:

> Thus, you see the world did not start with almighty Dicke.

The Big Bang theory had some problems: according to Hubble's original estimates for the age of the universe, the Earth was older than the universe

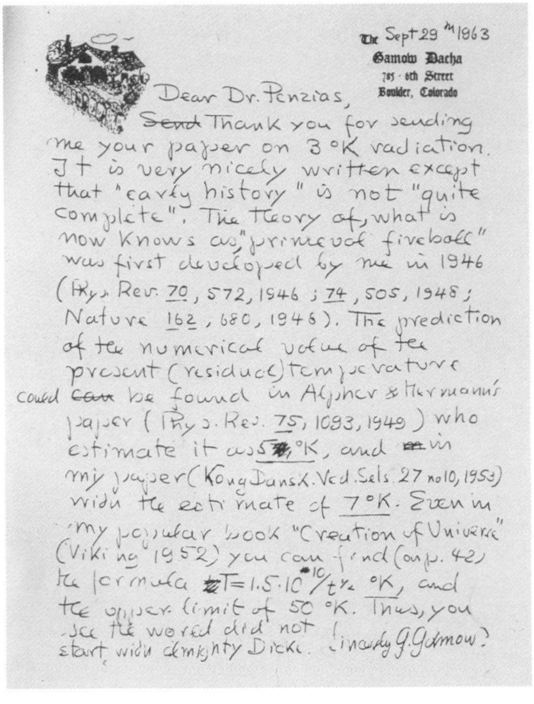

The Sept 29th 1963
Gamow Dacha
785 - 6th Street
Boulder, Colorado

Dear Dr. Penzias,

~~Send~~ Thank you for sending me your paper on 3 °K radiation. It is very nicely written except that "early history" is not "quite complete". The theory of what is now known as "primeval fireball" was first developed by me in 1946 (Phys. Rev. 70, 572, 1946; 74, 505, 1948; Nature 162, 680, 1948). The prediction of the numerical value of the present (residual) temperature could be found in Alpher & Hermann's paper (Phys. Rev. 75, 1093, 1949) who estimate it as 5 °K, and in my paper (Kong Dansk. Ved. Sels. 27 no 10, 1953) with the estimate of 7 °K. Even in my popular book "Creation of Universe" (Viking 1952) you can find (on p. 42) the formula $T = 1.5 \cdot 10^{10}/t^{1/2}$ °K, and the upper limit of 50 °K. Thus, you see the world did not start with almighty Dicke.

Sincerely G. Gamow

Figure 10.12 A letter from an annoyed George Gamow to Arno Penzias pointing out that there is even a discussion of the microwave background radiation predicted by the Big Bang – primeval fireball – in his popular book *Creation of the Universe*. The letter is misdated by two years. [Courtesy Elfriede Gamow.]

itself! It was partly because of such problems, and partly from aesthetics, that Herman Bondi, Thomas Gold and Fred Hoyle devised an alternative mechanism for the expansion of the universe. This involved the continuous creation of matter and yielded a universe that was continually expanding. The discovery of the background radiation, along with detailed examination of other evidence, have now ruled out this 'Steady State' theory of the universe. In 1967, the cosmologist Dennis Sciama wrote:

> … the loss of the Steady State theory has been a cause of great sadness. The Steady State theory has a sweep and beauty that for some unaccountable reason the architect of the universe appears to have overlooked. The universe in fact is a botched job.

Figure 10.13 The Cosmic Background Explorer satellite (COBE), launched in 1989, which discovered ripples in the cosmic background radiation. [NASA/Goddard Space Flight Center, Greenbelt, Maryland.]

More recently, measurements of the microwave background radiation made by the Cosmic Background Explorer satellite (COBE) have made headline news. After the euphoria of its discovery, it was realized that the apparent uniformity of the background radiation posed problems for the Big Bang

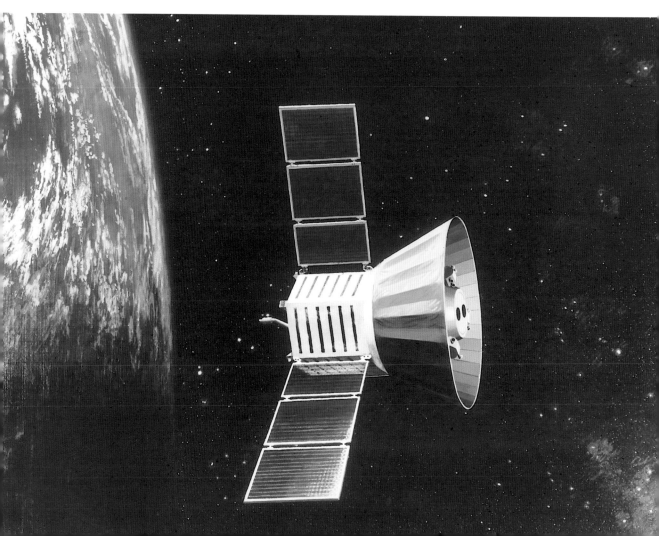

theory in terms of the apparent very short timescale for galaxy formation. The COBE data have revealed 'ripples' in the background sea of radiation that alleviate these problems.

Black holes and all that

Probably one of the best known ideas in modern astronomy is that of a 'black hole' – an object so massive that not even light can escape its attractions. The germ of this idea dates back to the work of John Michell and Pierre Simon Laplace in the eighteenth century. In their modern incarnation, such objects caught the public's imagination after John Wheeler coined the phrase 'black hole' at an astrophysics conference in New York in 1967. A black hole constitutes the most extreme end-point in the evolution of a star. It is the remnant of a collapsed star, whose gravitational field is so intense that it is surrounded by a region of space-time from which not even light can escape. The connection of such objects to Einstein's general relativity starts with a German astronomer named Karl Schwarzschild. In January and February of 1916, Einstein read two papers to the Prussian Academy on behalf of Schwarzschild, who was on active military service. The first paper contained an exact solution to the general relativistic equations for a static point mass: this is now known as the Schwarzschild solution. Although both Einstein and Schwarzschild viewed a 'point mass' as an unrealistic idealization, the solution contained the essence of a black hole. The point mass is surrounded by a region of space-time from which light is unable to escape. The boundary of this region is now called the 'Schwarzschild radius' or the 'event horizon'. Karl Schwarzschild died on the 11th of May 1916, of an illness contracted while on duty at the Russian front.

Figure 10.14 Karl Schwarzschild (1873–1916) in academic robes in Göttingen, Germany. He formulated his classic work on general relativity whilst serving on the Russian front during the First World War. His son Martin is also a distinguished astronomer. [Photograph by Robert Bein, courtesy AIP Emilio Segrè Visual Archives.]

The gravitational effect on curved space-time far away from the Schwarzschild radius around a black hole is exactly the same as that caused by a star of the same mass, and light passing far away from a black hole will be only slightly deflected. As the light path approaches the black hole, the deflection becomes more and more extreme. At a certain critical distance, the beam of light will be captured and will trace a circular orbit about the black hole. Inside what is called the 'photon sphere' (a sphere with a radius equal to the critical distance of approach), is the event horizon: once light, or anything else, goes closer to the black hole than this it can never escape. After the event horizon has been crossed, nothing can be communicated to the outside universe. So far, all this theory seems to be in the realm of science fiction. Are black holes real observable physical objects, or merely artifacts of our mathematical idealizations?

In order to understand how objects such as black holes could be formed, we must review the current understanding of stellar evolution. A star like our Sun is prevented from 'gravitational collapse' (in which the attractive gravitational forces within the star tend to make the star collapse in on itself) by the outward pressure generated by nuclear reactions in the interior. Near the end of a star's life, the nuclear fuel for these reactions becomes exhausted,

Figure 10.15 The nearest planetary nebula is the Helix nebula in the constellation of Aquarius, about 400 light years away. It was formed when a red giant star shed its outer shell. The material continues to glow because the hot core of the star is cooling and still emitting radiation. The star at the core will eventually become a white dwarf star. [© Anglo-Australian Observatory; photograph by David Malin.]

and the star begins to contract under gravity. The ultimate fate of the star is controlled simply by its mass. A not very massive star, such as the Sun, will shed some of its outer layers to form a so-called 'planetary nebula', leaving the remaining core to contract to form a 'white dwarf' star. A white dwarf star contains about the same mass as our Sun in a volume about the size of the Earth. Hence, the density of white-dwarf matter is about one million times more dense than normal Earth matter, and consists of nuclei wandering about in a sea of electrons. It is the resistance of these electrons to being squashed together that prevents further gravitational collapse. Stars more massive than the Sun can come to a more spectacular end. Subrahmanyan Chandrasekhar, the Indian-born astrophysicist, demonstrated that if the mass of the star is greater than about 1.4 solar masses – the so-called 'Chandrasekhar limit' – the reluctance of the electrons to be compressed is not strong enough to overcome the gravitational forces trying to cause further collapse. Consequently, the electrons are absorbed by protons to form neutrons in a weak nuclear reaction, and the final collapse of the core is very rapid and violent. The resulting spectacular stellar explosion was named a 'supernova' by Walter Baade and Fritz Zwicky in 1934. Most of the stellar material is hurled out into space, but Baade and Zwicky proposed that an exotic form of nuclear matter is left behind as a 'neutron star'. The density of

Figure 10.16 Fritz Zwicky (1898–1970) was a Swiss astronomer who worked for most of his life at the California Institute of Technology in Pasadena. He was among the first to propose that supernovae create neutron stars. Zwicky produced a major six-volume catalogue of galaxies and galaxy clusters, and showed that most galaxies are members of clusters. [Courtesy Caltech.]

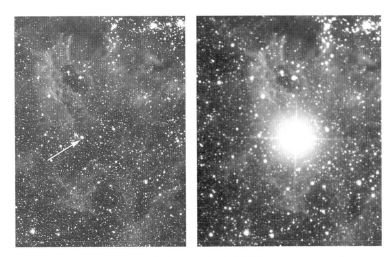

Figure 10.17 Supernova 1987a before (left) and after it exploded. This supernova occurred in one of the companion galaxies to the Milky Way and was visible to the naked eye. Although a large amount of energy was emitted in the visible spectrum, the vast bulk of its energy was emitted as neutrinos. Neutrino detector experiments on Earth were able to detect the supernova event. [© Anglo-Australian Observatory; photographs by David Malin.]

matter in such objects is such that a neutron star about twice as massive as our Sun would occupy a sphere of only 10 miles in diameter. The discovery of 'pulsars' in 1967, and their subsequent theoretical identification as rotating neutron stars, were important steps towards the acceptance of the probable existence of even stranger objects – black holes.

The modern theory of black holes begins with a 1939 paper by J. Robert Oppenheimer and Hartland Snyder entitled 'On continued gravitational contraction'. Their paper showed that if a star's collapsing core had sufficient mass, then the neutron pressure would be insufficient to prevent further collapse of a neutron star. They wrote:

> The star thus tends to close itself off from any communication with a distant observer; only its gravitational field persists.

An important feature of this calculation was the idea from general relativity that, as the star becomes more and more compressed, the pressure itself adds to the gravitational attraction, so accelerating the collapse. Despite this concrete suggestion, most physicists, Einstein included, were suspicious of the Schwarzschild 'singularity'* and did not take Oppenheimer and Snyder's suggestion very seriously.

The subject of black holes was resuscitated in the 1960s by a new theoretical result discovered by a mathematician from New Zealand and by the discovery of quasars. In 1963, Roy Kerr discovered a new exact solution

*A singularity is a value of parameters where an equation 'blows up', i.e. becomes infinite. A simple example is given by $1/x$, which has a singularity at $x=0$.

Figure 10.18 The Cambridge physicist Stephen Hawking. Hawking is the author of *A Brief History of Time*, probably the best-selling popular science book ever. [Courtesy Prof. S. Hawking.]

to Einstein's field equations. This solution is now recognized as the solution for a rotating black hole: the Schwarzschild solution is for the special case of zero rotation. Are these the only solutions? After many years of research, the problem was finally solved by Werner Israel, Brandon Carter and Stephen Hawking in a theorem known cryptically as 'black holes have no hair'. This summarizes the fact that whatever additional features, or 'hair', one tries to add to the collapsing core, the only things that ultimately matter are the mass and 'angular momentum' of rotation of the black hole. In the 1960s, Hawking, Roger Penrose and others also proved a number of important theorems about black holes. Their combined conclusion was that black holes are inevitable in any theory of gravity with an attractive force and in which Einstein's mass–energy equivalence holds good. It became possible to talk about the laws of black-hole physics.

An important part of this new work on black holes arose from a theorem proved by Hawking in 1972. This stated that the area of a black-hole event horizon can never decrease. At about this time, Jacob Bekenstein was a graduate student at Princeton and was discussing physics with John Wheeler. They were talking about the increase in entropy – the increased disorder, or loss of information – that occurred when Wheeler mixed a cold cup of tea with a hot cup of tea. In typical fashion, Wheeler remarked:

> The consequences of my crime, Jacob, echo down to the end of time. But if a black hole swims by, and I drop the teacups into it, I conceal from all the world the evidence of my crime. How remarkable!

Some months later, Bekenstein came back with this reply:

> You don't destroy entropy when you drop those teacups into the black hole. The black hole already has entropy, and you only increase it!

Bekenstein had noticed that Hawking's result for the area of the black hole was reminiscent of the second law of thermodynamics, which says that entropy can never decrease. When a particle is captured by a black hole, the mass and entropy of the black hole increases. Bekenstein connected this increase in entropy with an increase in area of the black hole. Hawking was at first sceptical of such an application of his ideas, but by 1974 he had become an enthusiastic supporter of the thermodynamic ideas of Bekenstein. In a cunning application of quantum mechanics to the space-time around black holes, he proposed that high-mass black holes have a low temperature and that low-mass ones have a high temperature. Hawking also suggested that these low-mass 'mini-black holes' could have been formed in the Big Bang and would eventually evaporate in spectacularly violent explosions. This theoretical analysis of the singularity structure of general relativity is mathematically very beautiful – but is there any experimental evidence for the existence of such astronomical objects as black holes?

Necessarily, all the observational evidence for black holes is indirect and therefore subject to some doubt and debate. A first problem lies in distinguishing black holes from neutron stars. A neutron star causes a considerable distortion of space-time in the vicinity of its surface, and matter falling towards either a neutron star or a black hole will be accelerated to speeds approaching the velocity of light. We therefore expect both neutron stars and black holes to be strong sources of X-rays – very high-energy photons – caused by the charged matter falling towards them. The search for black holes therefore begins with X-ray surveys of the sky. How do we distinguish between X-rays due to black holes and those from neutron stars? The key parameter is the mass of the object concerned. A collapsed object with a mass in excess of about three times the mass of the Sun cannot be a neutron star: it would be unstable against further collapse. But how do we measure the mass of such an object? The answer lies in concentrating on binary X-ray sources. As many as half the stars in our galaxy are believed to occur in binary systems, in which two stars are in orbit round each other. In optical binary systems, the motion of each star in orbit can be identified from the Doppler shifts in the light from the stars. The periods of such optical binary systems range from a few hours, for close binary systems, to hundreds of years. From this period of orbital rotation, astronomers can deduce information about the masses of the stars involved. Sometimes only one of the partner stars can be visually identified. If the mass of the observed star is known, the mass of the unobserved partner can be calculated from the orbital period. Can we use the same techniques for X-ray binary systems?

In December 1970, an X-ray satellite was launched off the coast of Kenya. It was called 'Uhuru', which is Swahili for 'freedom'. The hundreds of X-ray sources identified by Uhuru brought X-ray astronomy into the forefront of mainstream astronomy. One of the sources catalogued by Uhuru was called Cygnus X1 – the first identified X-ray source in the constellation of Cygnus. Measurements relating to the source were sufficiently accurate for an optical partner to this X-ray source to be identified: this was a bright blue star that appeared to be orbiting an unseen companion every five and a half days. Using the best estimate of the mass of this companion star enabled astronomers to tie down the mass of the X-ray source to be about six solar masses, and therefore much too large to be a neutron star. Although the evi-

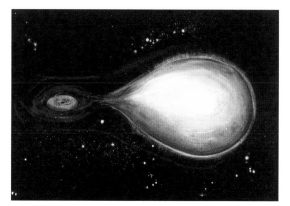

Figure 10.19 An artist's impression of the X-ray binary system Cygnus X1. A giant distorted star is losing mass to an orbiting black hole. The accumulated material orbits the black hole in an 'accretion disc' before eventually falling beyond the event horizon. [Griffith Observatory, Lois Cohen.]

dence is indirect, this source remains one of the best candidates for a black hole. Subsequently, other binary X-ray sources have been analysed and have identified other black hole candidates as well as many that are likely to be neutron stars. The general coherence of theory and experiment for neutron stars, black holes and gravitational collapse has convinced most physicists to take black holes seriously. The jury is still out for mini-black holes. There are doubts as to the mechanisms for their formation, and no astronomical evidence for a mini-black hole explosion exists as yet.

Figure 10.20 Maarten Schmidt is the Dutch-American astronomer who was the first to unravel the optical spectra of quasars. [Courtesy Caltech.]

The quasar problem

There remain many outstanding questions for physicists and astronomers studying black holes. The nature of quasars is still one of the most debated topics in astrophysics, and most astronomers now believe that supermassive black holes are a part of the answer. The first quasar was identified by Maarten Schmidt in 1963. He realized that the puzzling optical spectra from the object 3C 273 was due to its very large red-shift. It is now generally believed that the large red-shifts of quasars are evidence that they are very far away – large red-shifts mean large velocities, which correspond to large distances according to Hubble's law. Apart from the large red-shifts, the other remarkable feature of quasars is the huge amount of energy that they radiate, despite the fact that they occupy a very small region of space. At the time of its discovery, the star-like object 3C 273 was almost as distant as the most distant galaxies then known, yet it outshone our galaxy by about 1000 times. Nowadays, many thousands of quasars have been catalogued, and, although the first ones found were strong radio sources, these now constitute a minority.

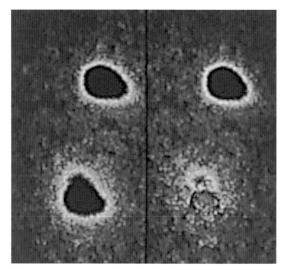

Figure 10.21 An optical false-colour image of the quasar 0957+561. On the left, two images of the quasar are seen with the image of the lensing galaxy merged with the lower quasar. On the right, the upper quasar image is subtracted from the lower image, revealing the lensing galaxy. [Institute for Astronomy, University of Hawaii, courtesy of Alan Stockton.]

One exciting property of quasars was the variation observed in their optical brightness. This is evident not only from recent experiments but also from old photographs which inadvertently included quasars. A good example is the quasar 3C 279, which flared up in brightness by a factor of about 25 in both 1936 and 1937. Some quasars have been seen to alter in intensity on timescales as short as a week or less. The significance of this variability is that it allows us to place a limit on the size of the power source at the heart of quasars. Suppose the source varies in intensity over a period of one week. Light travelling to us from the back of an object one light week in diameter arrives one week after light from the front. The fact that we see a variation in one week then means that the source must be less than one light week across. This distance is only about ten times the size of our solar system – for a source more powerful than one hundred galaxies! Quasars are one example of objects now known as 'active galactic nuclei', or AGNs. The most popular explanation for the enormous power generated by such objects is that it comes from a supermassive black hole, but the details of such theories are still somewhat speculative.

Two applications of general relativity

After some eighty years, during which general relativity has been subjected to more and more precise tests, it seems reasonable to assume that it is correct, and therefore that we can use the theory to uncover other new physics. In this section, we describe two such uses of general relativity: the first is connected with the bending of light; and the second involves the discovery of gravitational waves. We begin with the bending of light.

In 1979, sixty years after the first confirmation of the bending of starlight, an even more spectacular observation was made. This was the discovery of the 'double quasar' known as Q0957+561 (Figure 10.21). This consisted of optical and radio images of a pair of quasars separated by about 6 arcseconds. In itself, this would not have been so remarkable, except for the fact that the parameters of both quasars – the red-shifts and the spectra – were essentially identical. The astronomer Dennis Walsh, and his colleagues Robert Carswell and Roy Weymann, working at the University of Arizona and the Kitt Peak National Observatory, proposed that they were seeing two images of the same quasar, and that, somewhere along the line of sight, a massive object was deflecting light from the quasar in such a way as to produce a double image. Their suggestion was vindicated when Peter Young and co-workers using the 200 inch Hale Telescope at the Mount Palomar Observatory discovered a faint giant galaxy lying between the images. The red-shifts of the quasar and galaxy confirmed that the galaxy was indeed between us and the quasar.

The idea of such 'gravitational lensing' is not new. Arthur Eddington and Oliver Lodge both speculated on this possibility as early as 1919, and in the 1930s the question was taken up more seriously by the California-based Swiss astronomer Fritz Zwicky. In a paper in 1936, Einstein himself pointed out that the bending of light by massive bodies could have some strange

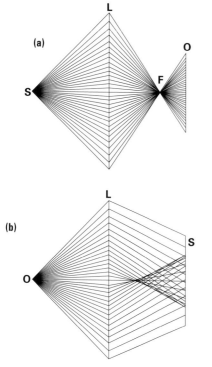

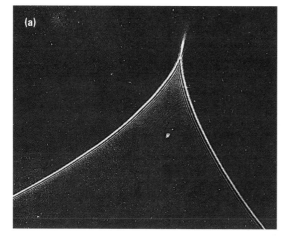

Figure 10.22 A diagram showing the difference between a linear lens system and a non-linear gravitational lens. In (a), light from a source (S) passes through a single lens (L), where it is deflected by an amount proportional to its displacement off the axis. The light then passes through a single focus (F) on its way to the observer at O. Only one ray from S is connected to any point at O. Part (b) shows a non-linear gravitational lens with rays traced backwards from a fixed observation point (O). The deflection now depends in a non-linear manner, on displacement off axis, and there is no longer a sharp focal point F; instead, a cusp caustic is formed. When a source S is located behind this cusp, three rays connect it to the observer and three images are seen. [Adapted from R. D. Blandford et al., *Science* (1989), vol. 245, p. 825.]

consequences, but he thought that the alignments required would be very improbable. By contrast, Fritz Zwicky suggested that gravitational lensing could have a major impact on cosmology, enabling us to 'weigh' nebulae and providing us with crude telescopes of interstellar dimensions with which to magnify the observed sources. It was over forty years before Zwicky's hopes began to be realized. There are now real prospects that gravitational lenses can help answer some of the outstanding puzzles of modern cosmology. Since the discovery of the first doubly imaged quasar, there has been secure observational evidence for six more multiply imaged quasars, as well as 'arc' and 'arclet' images of high-red-shift galaxies and ring images of extended radio sources.

How do such images come about? There are two important differences between gravitational lenses and conventional optical lenses. First, for thin optical lenses there is a linear dependence between the deflection angle and the distance of the ray from the axis of symmetry of the optical system. This means that there is a unique ray connecting a point source to each point in the plane of observation. By contrast, gravitational lenses do not have a simple linear dependence of the deflection with displacement. Such non-linear behaviour means that there can be more than one ray connecting a source to the observer: this is the origin of multiple images. The second difference between the two types of systems is that, instead of a

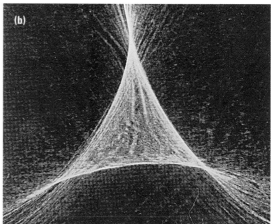

Figure 10.23 Although elementary optics talks about light rays passing through a focal point, more typically one finds that rays bunch together to form a caustic surface in space where the light intensity is greatest. The inner surface of an illuminated tea cup produces just such a caustic, whose intersection with the surface of the tea shows up as a cusp point, as in this picture. Part (a) shows two 'fold' caustics meeting at a 'cusp' point. Much more complex caustic patterns can be found, as shown in part (b); such caustics can be analysed using the so-called 'catastrophe optics' pioneered by Mike Berry from the University of Bristol in England. [Photographs by P. N. Kesterton and D. B. White, reproduced by courtesy of Professor J. F. Nye.]

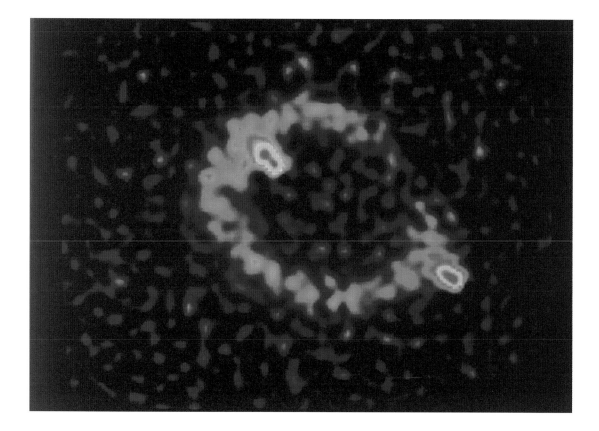

Figure 10.24 This radio image of an object called MG 1131+0456 represents a possible Einstein ring caused by the lensing galaxy being exactly on the line between the Earth and the radio source. [Courtesy NRAO/AUI; observations by J. N. Hewitt and E. L. Turner.]

fixed source of light, for gravitational lenses we have effectively a fixed observer and an extended source. The analysis of gravitational lenses is an application of geometrical optics. The familiar point focus of optical lenses is actually an idealization rarely realized by Nature. Much more common is a situation in which rays incident at different distances from the optical axis of the system are focussed to different points. Typically this results in a cusp-shaped curve of bright light known as a 'caustic'. Such cusp-like reflections may be seen on the surface of a cup of coffee in bright light. The caustic is caused by the curved inside surface of the cup acting as a mirror for the incoming light. More complex 'caustic networks' are often visible in the light patterns on the bottom of a swimming pool. Caustics and caustic networks play an important role in modelling gravitational lenses. Most examples can be approximated by assuming the lens matter to be smoothly distributed, but in some cases it is necessary to take account of individual stars in a lensing galaxy. The former situation is referred to as 'macrolensing', and the latter is called 'microlensing'.

The study of gravitational lensing is still in its infancy, but already many ingenious applications have been suggested. Assuming the correctness of Einstein's gravitational deflection, and of Shapiro's time delay for different light paths, much can be learnt about the distribution of matter in the universe. One of the outstanding puzzles in cosmology is the problem of 'dark matter'. It is now widely accepted that the observed distribution of stars can account for only a fraction of the mass in the universe. Observations of our galaxy, of spiral and elliptical galaxies, and of the velocities of galaxies within galactic clusters, can only be made consistent with the known laws of physics if there exists a large amount of additional matter that cannot be detected by its electromagnetic radiation, hence the name 'dark mattter'. With gravitational lenses, we can now use photons to probe the gravitational effects of intervening matter in much the same way as electron scattering is used to explore the atomic nucleus. There are now some intriguing results that may be due to the presence of dark matter in the universe.

Gravitational lenses are also helpful in 'cosmography' – the determination of the geometry of the universe on the largest scales. They have certainly confirmed our belief in the red-shift–distance relation for quasars, in that high-red-shift sources do appear to lie beyond lower-red-shift lenses. Attempts have also been made to make an accurate estimate of the Hubble constant by using measurements of the gravitational time delay. Uncertainties in the lens model mean that it is difficult to know what reliance to place on such results, although there is hope that radio ring-images may make it possible to constrain the lens model more precisely. A third astronomical use of a gravitational lens is one originally suggested by Zwicky, namely using the lens as a large telescope. In the spectacular examples of the blue arcs and radio rings, some parts of the source are magnified by quite large factors. It may also be possible to use microlensing as a fine resolution probe of quasars.

As a second application of general relativity, we turn to gravitational waves and the binary pulsar known as PRS 1913+16. Clifford Will has called this pulsar 'the first known system in which relativistic gravity can be used as a practical tool for the determination of astrophysical parameters'. How does this come about, and what has such a system got to do with gravitational waves? In Newton's theory of gravity, the interaction is supposed to be instantaneous. Einstein's gravity is a field theory, like that of electromagnetism, and, as in electromagnetism, disturbances or changes in the field do not propagate instantaneously but are limited by the speed of light. For electromagnetism, these disturbances are just light waves; therefore, for gravity we expect there to exist gravitational waves. One way of visualizing such waves is by using our representation of space as an elastic sheet with stars causing dents in the sheet. Imagine a binary star system, with the stars orbiting around their common centre of gravity. On our elastic sheet model of the universe, this can be represented as a pair of ball bearings moving round each other. As they move, ripples of distortions of the elastic sheet will propagate outwards from this disturbance: these ripples are a visualization of gravitational waves. If gravitational waves encounter a block of matter, the matter will experience a minute tremor. Einstein considered the possibility of such gravitational waves as early as 1916; however, instead of imagining a binary system, he calculated the energy emitted as gravitational waves by a rotating 'dumb-bell' – two balls of matter linked by a rod. Because the gravitational force is so much weaker than the electromagnetic force, the energy contained in these gravitational waves is expected to be tiny. Will, in his excellent book *Was Einstein Right?*, gives the following example. For one of the strongest imaginable sources of gravitational waves, a rotating star that collapses to form a black hole in our galaxy, the corresponding gravitational disturbance would cause two masses which are separated by 1 metre to move together and apart by only one-hundredth of the diameter of an atomic nucleus, or about 10^{-17} metres.

In 1968, Joseph Weber of the University of Maryland shocked the scientific world by his claim to have detected gravitational waves at an intensity over 1000 times larger than expected. Many other groups took up the challenge to build gravitational wave detectors, and it is now believed that Weber was mistaken. The search to observe gravitational waves directly is still in progress, but in 1978 Russel Hulse and Joe Taylor stole the show with exciting indirect confirmation of gravitational radiation. Hulse was a graduate student of Taylor who had the job of spending the summer of 1974 scanning the sky for signs of new pulsars using the Arecibo Radio Telescope in Puerto Rico. The graphic details of the actual moments of discovery of the binary pulsar system PRS 1913+16 are recounted in Will's book. Suffice it to say that Hulse and Taylor were persistent and perceptive enough to realize that the curious variations they had observed in the weak pulsar signal were significant. These variations were due to motion of the pulsar around its partner star. The data showed that the orbital period was about 7 hours 45 minutes, which explained the puzzling daily shift of 45 minutes in the pattern of period changes. From the Doppler shift of the signal they deduced that the orbital velocity of the pulsar varied from 75 kilometres/second to 300 kilometres/second. This maximum speed, about one-thousandth of the velocity of light, should be compared with the Earth's orbital speed of around 30 kilometres/second. Knowing both the speed and the orbital period of the pulsar enabled Hulse and Taylor to calculate that the circumference of its orbit is about equal to the circumference of the Sun. Thus, the pulsar must be separated from its companion star by only about one solar radius.

Figure 10.25 Joseph Taylor (right) with Robert Dicke. Taylor shared the 1993 Nobel prize for physics with his former graduate student Russell Hulse. Taylor and his wife invited Joycelyn Bell, discoverer of the first pulsar and considered by many to have been unlucky not to have won the prize herself, to be their guest at the ceremonies. [Photograph by Robert P. Matthews, Princeton University.]

Figure 10.26 The 1000 foot diameter radio telescope at Arecibo in Puerto Rico was constructed in a natural hollow in the ground. It is the largest dish antenna in the world, and it was used by graduate student, Russell Hulse, and his professor, Joseph Taylor, to discover the binary pulsar PSR 1913+16. The telescope was used as the location for the climax of the James Bond movie *Goldeneye*. [The Arecibo Observatory is part of the National Astronomy and Ionosphere Center, which is operated by Cornell University under a cooperative agreement with the National Science Foundation.]

Figure 10.27 Joseph Weber with his gravitational radiation detector. The detector is a large aluminium cylinder with 'piezoelectric strain' crystals glued to its surface. These crystals allow the detection of minute fluctuations in shape due to the passage of gravitational waves. [Courtesy Joseph Weber.]

Such a system is expected to show general relativistic corrections to the orbit at a level more than twenty times as large as the perihelion shift for Mercury. For a binary system the distance of closest approach is called the 'periastron'. In December of 1974, Taylor and Hulse measured the rate of advance of the periastron to be about 4 degrees/year. This is some 36 000 times larger than the perihelion advance of Mercury due to a combination of its larger intrinsic size and the fact that the pulsar completes many more orbits per year. Assuming the correctness of general relativity, we can estimate the total mass of the double star system. This number has now been very accurately determined to be 2.8275 solar masses. The binary pulsar PSR 1913+16 was unusual in another respect, and it had some more surprises in store. Pulsars are known to have pulse periods that are very steady but which, on detailed examination, may be seen to be very gradually slowing down. After removing the Doppler motion effects from the data, Hulse and Taylor's pulsar can be seen to be slowing down at an incredibly slow rate – only by about 4% in one million years. This extraordinary constancy enabled Taylor and his colleagues to make a very detailed analysis of the variations in the pulsar's period. In addition to the ordinary Doppler motion effect, there should be small corrections due to time dilation and the gravitational redshift. Assuming the correctness of Einstein's theory then enables us to determine the relative masses of the two stars. The calculations reveal that the two stars are very nearly equal in mass. The most probable candidate for the companion of the pulsar is another neutron star with its pulsar beam presumably pointing away from the Earth.

What has this to do with gravitational waves? The rotating binary star system is essentially equivalent to Einstein's dumb-bell radiator: we expect the binary pulsar to emit gravitational radiation. Because of the distance of the pulsar from the Earth, the predicted radiation is too weak to be

Figure 10.28 Gravitational wave detector at the California Institute of Technology in Pasadena. This experiment uses a system of laser beams to detect the tiny motions caused by gravitational waves and is sensitive to changes smaller than the size of a proton. This apparatus is a prototype for a much larger gravitational wave observatory planned for the near future. [LIGO project, Caltech.]

detected directly, but the system is nonetheless radiating energy in these gravitational waves. This gravitational energy loss is predicted to show up as a tiny decrease in the orbital period of the pulsar – a mere seventy-five-millionths of a second per year for a period of some 27 000 seconds. In December of 1978, just in time to celebrate the centenary of Einstein's birth in 1879, Joe Taylor was able to announce agreement with Einstein's prediction. Five years later, new data taken by Taylor were able to improve the accuracy of this measurement. The Einstein prediction now agrees with experiment to within 3%. This result is a stunning success for general relativity, and it strengthens our confidence in using the theory as a tool to explore other new phenomena. The hunt for direct observations of gravitational waves has now intensified using a new generation of more sensitive detectors.

The search for a unified theory

I don't like that they're not calculating anything. I don't like that they don't check their ideas. I don't like that for anything that disagrees with an experiment, they cook up an explanation – a fix-up to say 'Well, it still might be true.' For example, the [superstring] theory requires ten dimensions. Well, maybe there's a way of wrapping up six of the dimensions. Yes, that's possible mathematically, but why not seven? When they write their equation, the equation should decide how many of these things get wrapped up, not the desire to agree with experiment. In other words, there's no reason whatsoever in superstring theory that it isn't eight of the ten dimensions that get wrapped up and that the result is only two dimensions, which would be completely in disagreement with experience. So the fact that it

might disagree with experience is very tenuous, it doesn't produce any-
thing; it has to be excused most of the time. It doesn't look right.
Richard Feynman, in *Superstrings: A Theory of Everything?*

Einstein spent much of the last thirty years of his life searching unsuccessfully for a unified field theory of electromagnetism and gravity. At the time he
began his quest, only these two forces had positively been identified, and it
was an intriguing question as to whether or not there was a deep underlying
connection between these two 'inverse square' laws of physics. Einstein's first
paper on unified field theory was published in 1922. It was only in 1921 that
James Chadwick had first shown that, at very small distances, the interactions
of alpha particles with an atomic nucleus did not follow exactly the inverse
square law predicted from the repulsion of their positive electric charges.
Chadwick concluded that his experiments showed that these nuclear forces
are of 'very great intensity'. According to Pais, this is the first published statement about the existence of a strong nuclear force. This 'new force' interpretation was disputed until well into the 1920s. The neutron was discovered by
Chadwick in 1932, and this discovery led to our modern picture of a nucleus
made up of neutrons and protons held together by the 'strong interactions'.
Enrico Fermi published his famous theory of beta decay and weak interactions in 1934, but only after his paper had earlier been rejected by the prestigious journal *Nature*. In 1922, when Einstein published his first paper on
unified field theory, electrons and protons were thought to be the only fundamental constituents of matter, and the only known forces were electromagnetism and gravity. Why did Einstein not modify his goal of unifying only the
electromagnetic and gravitational fields after the discovery of the weak and
strong nuclear forces? No one can say, but his pursuit of this 'holy grail' of a
unified field, and his distrust of the very quantum theory that he had done so
much to bring about, isolated Einstein from the mainstream of nuclear and
particle physics for the rest of his life. In 1942, he wrote to a friend:

I have become a lonely old chap who is mainly known because he does not
wear socks and who is exhibited as a curiosum on special occasions.

But in the same letter he also wrote:

In regard to work, I am more fanatic than ever and really hope to have solved
my old problem of the unity of the physical field. However, it is like having
an airship with which one can sail around in the clouds without seeing clearly how to land in reality, that is, on earth.

We now know that the weak and electromagnetic forces can indeed be
described by a unified field theory. The resulting 'electroweak theory',
together with the 'quark' symmetry that describes the strong forces, is now
known as the 'standard model'. There are also some hopes that the strong
forces can be unified with electroweak forces in a true 'Grand Unified
Theory'. Sheldon Glashow, author of one of the original grand unified mod-

els, characterized its most striking prediction – that the proton should occasionally decay – as the problem of finding out whether or not 'a diamond is forever'. The non-observation of proton decay at rates predicted by these original models, together with some detailed problems in the formulation of such theories, have combined to dampen much of the early enthusiasm of the early 1980s. Much more noise and excitement have been generated by the possibility of a 'Theory of Everything'. This is a modern incarnation of Einstein's dream of a unified field theory incorporating gravity. These recent developments go under the name of 'string theories' – but, as Richard Feynman argues in the quotation at the beginning of this section, such theories have yet to demonstrate any predictive power. String theories incorporate, albeit in a different context, some of the ideas that Einstein pursued in his thirty-year quest.

We begin our brief survey of unified field theory by introducing Emmy Noether. Noether was the first woman to be appointed to a university position in Göttingen, just after the turn of the century. There was great opposition to her appointment from many members of the all-male faculty, but, fortunately, the famous mathematician David Hilbert was in favour of her appointment. With the arrival of the Nazis, Noether eventually had to flee from Germany. When she died in 1935, Einstein wrote:

> In the judgement of the most competent living mathematicians, Fraulein Noether was the most significant creative mathematical genius since the higher education of women began.

Her work, published in 1918, on the connection between symmetries and conservation laws is known as 'Noether's theorem', and it is a cornerstone of modern field theory. An example will give some idea of the power of her theorem. The fact that physics does not depend on the origin from which we choose to measure distances is an example of a symmetry. This symmetry is called 'translational invariance', and the corresponding conservation law is the conservation of momentum.

In 1919, a German mathematician named Theodor Kaluza wrote to Einstein suggesting that electromagnetism was a consequence of a symmetry involving a new, hidden fifth dimension. The fifth dimension was supposed to be curled up into a very small circle and be effectively unobservable. Writing Einstein's equations with this fifth dimension gives an extra set of equations which Kaluza identified with Maxwell's equations for the electromagnetic field. Charge conservation then arises as momentum conservation in this fifth direction. Einstein replied:

> The idea of achieving [a unified theory] by means of a five-dimensional cylinder world never dawned on me. … At first glance I like your idea enormously.

Osker Klein presented an improved version of Kaluza's ideas in 1926, and suggested that the circular nature of the fifth dimension was the origin of

Figure 10.29 Osker Klein (1894–1977) attempted one of the earliest 'Theories of Everything' combining gravity, electromagnetism and quantum physics. His ideas were based on an earlier extension of general relativity to five dimensions by Theodor Kaluza. The extra dimension incorporated Maxwell's equations as well as Einstein's gravity. [AIP Emilio Segrè Visual Archives.]

charge 'quantization' – the fact that all electric charges appear to be exact multiples of the charge on the electron. Abraham Pais writes in his book *Subtle is the Lord* that George Uhlenbeck said of this time:

> I remember that in the summer of 1926, when Osker Klein told us his ideas which would not only unify the Maxwell with the Einstein equations but also bring in the quantum theory, I felt a kind of ecstasy! Now one understands the world!

Einstein continued to explore the implications of such a 'Kaluza–Klein theory' for the next twenty years. Einstein also hoped that unification would be able to resolve what he perceived as limitations of quantum theory. In 1954, a year before his death, he said:

> I must seem like an ostrich who forever buries its head in the relativistic sand in order not to face the evil quanta.

With the discovery of the weak and strong forces, the 'Kaluza–Klein' theory became relegated to the status of a mathematical curiosity. The discovery of the myriads of new 'elementary' particles in the 1950s and 1960s seemed to rule out any hope of a simple unified field theory. However, the picture was dramatically simplified by the arrival of Murray Gell-Mann and George Zweig's 'quark model' of elementary particles. This was followed by independent suggestions from Sheldon Glashow, Abdus Salam and Steven Weinberg for a unified model of the weak and electromagnetic interactions. This unification is brought about at the price of introducing a mysterious new particle, called the 'Higgs' particle after its inventor, Peter Higgs from Edinburgh University. Lewis Ryder, a graduate student of Higgs, remembers the discovery of the 'Higgs mechanism' well. He came back to the department after a weekend hiking in the Scottish mountains to find a note on his desk from Higgs saying 'I have had my first good idea in ten years!' The key to unifying these interactions was a symmetry known as 'gauge invariance'. A better name for this symmetry would be 'phase invariance', since the symmetry arises because of the freedom for all the matter fields in these theories to be multiplied by an arbitrary space-time dependent phase with no change in the resulting physics. Gravity may also be understood as a type of 'gauge theory', as can the quark field theory – quantum chromodynamics or QCD – that is now believed to describe the strong interactions. The fact that all these theories are thought to be different varieties of gauge theories makes it look as if a total unification of all the forces may be possible after all. But there is a further complication. Both the electroweak theory and QCD are found to be consistent with quantum mechanics. In the early days of quantum field theory, calculations were plagued with 'infinities' – places where divergent integrals occurred. It was for providing a consistent calculational framework for quantum electrodynamics (QED) – the 'renormalization programme' – that Richard Feynman, Julian Schwinger and Sin-itiro Tomonaga were awarded the Nobel Prize in 1965. It was the Dutch physicist Gerard 't Hooft, while a graduate student of Tini Veltman in Utrecht, who first showed how these 'renormalization' techniques could be extended to the electroweak theory and QCD. The problem is that these techniques appear to fail when one tries to construct a quantum field theory of gravity. This apparent incompatibility of gravity and quantum mechanics remains one of the outstanding problems of theoretical physics.

The appeal of 'string theories' is that they represent a serious attempt to overcome this inconsistency between gravity and quantum mechanics. Edward Witten, a leading advocate of these theories, went so far as to say that for theoretical physicists not to work on string theory was equivalent to physicists in the 1920s not working on quantum mechanics! String theories combine a number of old and new ideas. The first is the extension of the number of dimensions to numbers like ten or twenty-six. The second is a curious symmetry known as 'supersymmetry'. The existence of such a supersymmetry would be signalled by the discovery of so-called 'supersymmetric partners' for all the known elementary particles. Besides the photon, there should be a 'photino'; besides the quarks there should be 'squarks'; and so on. So far, there is no experimental evidence for such super-

Figure 10.30 Mike Green, now Professor of Physics at the University of Cambridge (left), and John Schwarz, Professor at the California Institute of Technology in Pasadena (right), made a key breakthrough in 1984 in understanding how to formulate a self-consistent superstring theory. Both Schwarz and Green had worked on string theory since its first introduction by the Italian physicist Gabrielle Veneziano in the late 1960s. Unlike many others, they both had faith that string theory was 'so magical that it would get around the anomaly problem' which appeared to beset the theory. For their anomaly calculation to work required many peculiar looking numbers to add up to 496 – and they did! Curiously enough, given Feynman's views on string theory, the answer to the question 'Who got Feynman's office?' is John Schwarz. [Photograph of Green by Sharon Kinder-Geiger. Photograph of Schwarz courtesy John Schwarz.]

symmetric doubling. The last ingredient is the idea of a vibrating 'string' to replace the point particles of standard field theories. A superstring can have different types of vibration, and the lowest mode of each sort of vibration then corresponds to the electron, the photon, the neutrino, the 'graviton', and all the other elementary particles. The appeal of such a theory is obvious, but the realization of a superstring theory with any predictive power for the particles we can observe has not even begun. String theory predicts new effects corresponding to harmonics of the lowest modes of vibration. These are predicted to occur at excitation energies that are multiples of the so-called 'Planck mass'. This mass scale is so much larger than ordinary mass scales in particle physics – over one million million million times heavier than the mass of the proton – that it is extremely unlikely that any direct test of these predictions will ever be feasible. Compared with the scale of the Planck mass, all the presently known elementary particles have zero mass to a good approximation: the actual masses that we observe are presumed to result from, as yet uncalculated, 'symmetry breaking corrections'.

The debate about unification and string theories continues to excite particle physicists and astrophysicists. Great physicists on both sides of the debate have made sweeping statements concerning both physics and philosophy. By way of conclusion, we shall quote remarks made by two of the leading protagonists, Edward Witten and Sheldon Glashow. Which side of the divide each is on is evident from their remarks.

Figure 10.31 Edward Witten is a Professor at the Institute for Advanced Study in Princeton, New Jersey, where Einstein spent the later years of his life. He is one of the most lucid and persuasive advocates for superstring theory, and has made many important contributions to our understanding of both the physics and the mathematics of superstrings. Although Witten did not get Einstein's office, Princeton is a fitting place for him to pursue an updated version of Einstein's search for a unified theory. [Courtesy Prof. E. Witten.]

Edward Witten: Einstein developed general relativity at a time when the basic ideas in geometry that he needed had already been developed in the nineteenth century. It's been said that string theory is part of the physics of the twenty-first century that fell by chance into the twentieth century. That's a remark that was made by a leading physicist about fifteen years ago. What he meant was that human beings on planet Earth never had the conceptual framework that would lead them to invent string theory on purpose. No one invented it on purpose, it was invented in a lucky accident. By rights, twentieth century physicists shouldn't have had the privilege of studying this theory. What should have happened, by rights, is that the correct mathematical structures should have been developed in the twenty-first or twenty-second century, and then finally physicists should have invented string theory as a physical theory that is made possible by those structures. If that had happened, then the first physicists working with string theory would have known what they were doing perhaps, just like Einstein knew what he was doing when he invented general relativity. That would have perhaps been a normal way for things to happen but it wouldn't have given twentieth century physicists the chance to work on this fascinating theory. As it is we have had the stroke of good luck that string theory was invented in a sense without out human beings on planet Earth really deserving it.

Sheldon Glashow: I'm particularly annoyed with my friends, the string theorists, because they cannot say anything about the physical world. Some of them are convinced in the uniqueness and beauty, and therefore truth, of their theory, and since it is unique and true it obviously includes a description of the entire physical world. It does not seem to be necessary to do any experiments to prove such a self-evident truth, so they begin to attack the value of experiments from this end – a highly theoretical, abstract, mathematical end – whereas some of our friends in Britain are attacking physics from the other end, from the purely financial end. [They undermine the motivation for experiment] in the same way that I think medieval theology destroyed science in Europe in the middle ages. It was, after all, only in Europe that people did not see the great supernova of 1054, for they were too busy arguing how many angels could dance on the head of a pin!

Figure 10.32 Roger Penrose had his interest aroused in general relativity by Hermann Bondi and Denis Sciama. Besides his pioneering work on singularity theory, he has written a best-selling popular book called *The Emperor's New Mind*. The book contains a brilliant exposition of modern physics and computer science and attempts to rebut the concept of artificial intelligence. [Photograph by Steve Green.]

We have now come almost to the end of our story and, as in all stories, there are some loose ends. The most tantalizing loose end is the search for a consistent way to unify the two great pillars of modern physics – quantum mechanics and relativity. Einstein was deeply unhappy about the statistical uncertainty that is at the heart of quantum mechanics. Indeed, he hoped that his unification programme could remove this quantum uncertainty he found so abhorrent. Although string theory claims to produce a consistent theory of quantum gravity, it does not attempt to address the foundations of quantum mechanics. Roger Penrose, in his fascinating book *The Emperor's New Mind*, puts forward a very different thesis. He suggests that the measurement problem at the heart of quantum mechanics requires a radically new approach. Penrose believes, like Einstein, that general relativity still has a key role to play but at present, he has to admit to having 'only the germ of an idea' for the correct theory of quantum gravity.

How do we conclude? Let us take our leave with two final comments. First, a remark about Einstein by Feynman. In a conversation with Freeman Dyson, Feynman surmised that Einstein's great work had sprung from physical intuition, and that it was when 'he stopped thinking in concrete physical images and became a manipulator of equations' that Einstein's creativity ceased. And a final word from Einstein. In 1923, Einstein wrote to a colleague about unified field theory:

> Above stands the marble smile of implacable Nature which has endowed us more with longing than with intellectual capacity.

11   Afterword: Relativity and science fiction

The beginnings

The term 'science fiction' was first used by one of the founders of the modern genre, Hugo Gernsback. Gernsback, after whom the annual science fiction 'Hugo' awards are named, was the founder of the *Amazing Stories* magazine in April, 1926. The slogan on the title page proclaimed its mission: 'Extravagant Fiction Today … Cold Fact Tomorrow'. Of course, few of the stories published in *Amazing Stories* lived up to this claim, but science fiction does have some notable successes to its credit over its relatively brief history. Two of the founding fathers of science fiction – H. G. Wells and Isaac Asimov – have already been mentioned in earlier chapters. In this final chapter, we shall examine the interplay between relativity and science fiction. We begin with Johannes Kepler, arguably the first writer of the genre.

Kepler was born in south-west Germany in a small town called Weil-der-Stadt. Kepler's first great work, *A New Astronomy*, was published in 1609, and it remains a landmark in the history of science. In it, Kepler formulated the first 'natural laws' – precise, verifiable statements about natural phenomena expressed in terms of mathematical equations. Arthur Koestler, in his marvellous book *The Sleepwalkers*, claims that it was Kepler's laws that 'divorced astronomy from theology and married astronomy to physics'. Unlike Copernicus, Galileo or Newton, Kepler did not attempt to disguise the way in which he arrived at his conclusions – all his errors and sidetracks are faithfully recorded along with his final revelation. In a modern context, one can find a similar contrast between the traditional, formal approach of

Julian Schwinger and the informal, anecdotal account of Richard Feynman in the lectures they gave when they received the Nobel Prize for their (independent) work on quantum electrodynamics. Besides the scientific work for which Kepler is famous, one of his favourite projects was *Somnium* – an account of an imagined journey to the Moon.

Somnium was published posthumously in 1634, and is undoubtedly one of the forerunners of modern science fiction. The story is about a boy called Duracotus and his mother, Fiolxhilda, who lives in Iceland. Fiolxhilda is a witch who sells herbs and conversed with demons: interestingly, Kepler's own mother was accused of witchcraft and only narrowly escaped being burned at the stake. In the story, Duracotus is sold by his mother to a sea-captain, who leaves him on the Isle of Hveen, where he studies astronomy with the great Tycho Brahe. When he returns home five years later, his mother is repentant and summons up a friendly demon from the Moon. The demon is able to take ordinary mortals to the Moon during an eclipse, and the journey is described with a serious attempt at scientific accuracy:

> The initial shock is the worst part of it, for he is thrown upwards as if by an explosion of gunpowder. Therefore he must be dazed by opiates beforehand; his limbs must be carefully protected so that they are not torn from him and the recoil is spread over all parts of his body. Then he will meet new difficulties: immense cold and inhibited respiration. When the first of the journey is completed, it becomes easier because on such a long journey the

body no doubt escapes the magnetic force of the Earth and enters that of the Moon, so the latter gets the upper hand.

This extract also shows how close Kepler came to Newton's idea of universal gravity. After completion of the journey, Kepler describes conditions on the Moon: what he calls the 'Subvolvan' half always facing the Earth and the 'Prevolvan' half facing away; the fortnight-long, freezing lunar nights followed by equally long, scorching lunar days; the lunar mountains; and the seas. He also describes a genuinely 'lunatic' astronomy – the Sun, stars and planets as seen from the Moon. All this was firmly based on the astronomical knowledge of the time. Kepler goes on to imagine strange inhabitants of the Moon with cities surrounded by the circular walls of lunar craters. Kepler's footnotes make clear that part of his purpose in writing *Somnium* was an 'argument for the motion of the Earth, or rather a refutation of arguments constructed, on the basis of perception, against the motion of the Earth'. In another footnote, Kepler explains that the story is an allegory, with Duracotus representing Science and his mother symbolizing Ignorance. *Somnium* contained the essence of modern science fiction – an intriguing blend of science fact with imagination.

If Kepler is the founder both of modern science and of science fiction, then the Galileo and Newton of science fiction are undoubtedly Jules Verne and H. G. Wells. Jules Verne was born in Nantes, France, in 1828. After an undistinguished early career, Verne hit upon the idea of combining science with fiction to generate a new form of adventure story. In 1863, after finishing his first attempt, '*Five Weeks in a Balloon*', he wrote:

> I have just finished a novel in a new form, a new form – do you understand? If it succeeds, it will be a goldmine.

The book did succeed, and Verne went on to work his new goldmine. His novels were described as 'voyages extraordinaires' by his publisher, and they combined great attention to scientific detail with exciting stories and locations. In 1865, in *From the Earth to the Moon*, Verne described the construction of a gigantic cannon – the Columbiad – by the Gun Club in Baltimore. In a trial launch 'to ascertain the shock of departure' on life-forms, a cat and a squirrel were shot upwards on a parabolic trajectory to splash down in the sea. On opening the capsule, the cat leapt out, only slightly bruised, but there was no sign of the squirrel. Further investigation revealed that the cat had eaten the unfortunate squirrel. In reality, of course, the enormous acceleration from the cannon would have killed all the occupants of the shell. Yet in the narrative, with all its convincing scientific detail, the reader is persuaded to suspend disbelief. The description of the launch could almost be of a lift-off of the shuttle from Cape Canaveral:

> At the moment when that pyramid of fire rose to a prodigious height into the air the glare of the flame lit up the whole of Florida; and for a moment day superseded night over a vast expanse of the country. This immense canopy

Figure 11.2 Jules Verne (1828–1905) was one of the founders of modern science fiction. His father was a lawyer, but Verne was greatly influenced by meeting the French explorer Jacques Arago. At Arago's house he met many scientists and explorers, and this experience is woven into the background of many of Verne's novels. It was not until 1863 that Verne published his first 'voyage extraordinaire' – *Five Weeks in a Balloon* – the story of a crossing of Africa from Zanzibar to Senegal in a balloon-type aircraft. By the time of his death in 1905, Verne had written around sixty books including *Journey to the Centre of the Earth*, *20,000 Leagues under the Sea* and *Around the World in 80 Days*. [The Hulton Deutsch Collection.]

of fire was perceived at a distance of 100 miles out at sea, and more than one ship's captain entered in his log the appearance of this gigantic meteor. The discharge of the Columbiad was accompanied by an earthquake. Florida was shaken to its very depths. The gases from the gun-cotton, expanded by heat, forced the atmosphere back with tremendous violence, and this artificial hurricane rushed like a waterspout through the air.

The first astronauts had only a one-way ticket to the Moon: the novel ends with their faithful Earth-bound friend observing the new satellite of the Moon through a giant telescope. Verne's novel was prophetic in two ways: he foresaw the splashdown technique for the recovery of the early space capsules, and he also described the construction of the giant telescope located on top of Long's Peak in the Rockies. The 100 inch telescope on Mount Wilson in California was completed some fifty years after Verne's book was published.

These early stories grappled with the effects of Newtonian gravity and Newton's laws of motion. The writer who definitively enlarged the stage on which science fiction could take place was H. G. Wells. Herbert George Wells was born in Bromley, England, in 1866. The period in the nineteenth century surrounding Wells' birth had seen many startling scientific discoveries: Darwin published his *Origin of Species* in 1859; by 1864 Maxwell had unified electricity and magnetism, and shown that light was an electromagnetic wave; Mendeleev brought some order to chemistry with his Periodic Table of the Elements in 1869; Joule discovered the relationship between heat and work in 1847; and Lord Kelvin and others were laying the foundations of the theory of thermodynamics. It was not surprising therefore that science and technology were seen to be investments for the future. Coming from a poor family, Wells was awarded a scholarship (of one guinea a week) to attend the Normal School of Science in Kensington. It was at this school that Wells was inspired by the lectures of T. H. Huxley, a leading scientist of the day and a redoubtable defender of Darwin's theory of evolution by natural selection. After spending some years as a teacher, in 1890 Wells obtained a first class degree in zoology from London University. During forced inactivity after a football injury, he began to write. In 1893, Wells wrote a textbook of biology; in 1895, his novel *The Time Machine* was published. It was this novel that established Wells' reputation as a writer with extraordinary powers of imagination and invention. *The Time Machine* was written ten years before Einstein published the special theory of relativity. At the beginning of the book, the Time Traveller (as Wells' narrator refers to him) explains the idea of the fourth dimension to his friends:

> There are really four dimensions, three of which we call the three planes of Space, and a fourth, Time. There is, however, a tendency to draw an unreal distinction between the former three dimensions and the latter, because it happens that our consciousness moves intermittently in one direction along the latter from the beginning to the end of our lives.

In the ensuing discussion, the Psychologist argues that 'you can move about in all directions in Space, but you cannot move about in Time'. The Time Traveller replies:

> That is the germ of my great discovery. But you are wrong to say we cannot move about in Time. Of course we have no means of staying back for any length of Time, any more than a savage or an animal has of staying six feet above the ground. But a civilized man is better off than the savage in this respect. He can go up against gravitation in a balloon, and why should he not hope that ultimately he may be able to stop or accelerate his drift along the Time-Dimension, or even turn about and travel the other way?

Figure 11.3 Herbert George Wells (1866–1946), better known as H. G. Wells, is one of the giants of science fiction, who, with Jules Verne, helped to create the modern genre. He studied zoology at London University, and in 1895, after some years teaching, Wells published a 'scientific romance' entitled *The Time Machine*. This novel catapulted Wells to fame as a writer and opened up a whole new perspective for science fiction. In the next few years he wrote some of his most famous novels – *The Island of Dr. Moreau*, *The Invisible Man*, *The War of the Worlds* and *The First Men in the Moon*. It was with a radio dramatization of *The War of the Worlds* in 1938 that Orson Welles almost caused a riot in New York. Wells is often credited with forseeing the development of the tank, of aircraft and air warfare, of the atomic bomb and the nuclear stalemate that would inevitably follow, and even of a form of genetic engineering. Wells supported the First World War, and for a short time in 1918 he was the Director of Propaganda Policy against Germany. After the war, he vigorously promoted his view that the human race must work together for a World State or face extinction. He was deeply depressed by the Second World War; he died shortly after the publication of his last despairing work *Mind at the End of its Tether*. [The Hulton Deutsche Collection.]

In the remainder of the book, Wells has his Time Traveller exploring a rather bleak future, and the use of the time dimension in the novel has little to do with relativity. Verne and Wells wrote very different types of science fiction: Verne preferred to stay close to modern scientific reality, whereas Wells was not afraid to make leaps of the imagination. The contrast is nowhere better seen than in their novels about a voyage to the Moon. As we have seen, Verne envisaged using an extrapolation of existing technology in the form of a giant gun to launch the spaceship. Wells, on the other hand, invented a new material, 'Cavorite', that was opaque to gravity, and his spaceship was made of a bubble of Cavorite.

> He pointed to the loose cases and bundles that had been lying on the blankets in the bottom of the sphere. I was astonished to see that they were floating now nearly a foot from the spherical wall. Then I saw from his shadow that Cavor was no longer leaning against the glass. I thrust out my hand behind me, and found that I too was suspended in space, clear of the glass.
>
> I did not cry out or gesticulate, but fear came upon me. It was like being held and lifted by something – you know not what. The mere touch of my hand against the glass moved me rapidly. I understood what had happened, but that did not prevent me being afraid. We were cut off from all exterior gravitation, only the attraction of objects within our sphere had effect. Consequently everything that was not fixed to the glass was falling – slowly because of the slightness of our masses – towards the centre of gravity of our little world, which seemed to be somewhere about the middle of the sphere, but rather nearer to myself than Cavor, on account of my greater weight.

At the end of this chapter we shall see that there is now a new incarnation of this idea of 'anti-gravity' propulsion.

In his later novels, Wells is credited with several successful predictions. In *The Argonauts of the Air*, in 1895, he not only predicted manned flight by propellor-driven aircraft, but also foresaw the inevitability of the

> flying-machine that will start off some fine day, driven by neat 'little levers', with a nice open deck like a liner, and all loaded up with bomb-shells and guns.

His scientific background also enabled him to keep abreast of new scientific discoveries. From reading Frederick Soddy's 1909 book *The Interpretation of Radium*, Wells conceived the idea of a chain reaction and an atomic bomb, and as we have already seen, he wrote about this in his prophetic book *The World Set Free*. As with his flying machine, Wells took the next step and foresaw the inevitable nuclear stalemate that would result, with weapons too terrible to use and wars that no one could win. This ability of science fiction to look two steps ahead has been memorably summarized by Fred Pohl, a modern science fiction writer:

> A good science fiction story should be able to predict not the automobile but the traffic jam.

Wells made many lasting contributions to science fiction, but it was his popularization of time as a fourth dimension which liberated writers from mere spatial explorations and encouraged them to think of the far future.

The 'Golden Age'

The first quarter of the twentieth century saw momentous changes in our understanding of physics. In earlier chapters, we have described the discovery of radioactivity and the birth of nuclear physics. In ten remarkable years, Einstein transformed our understanding of space and time, of mass and energy, and of gravity and acceleration. Atomic theory progressed from Rutherford's nucleus and Bohr's atom to relativistic quantum mechanics and anti-matter. With the completion of the 100 inch telescope at the Mount Wilson Observatory, the understanding of the universe also made great strides. The telescope became operational in 1917; one year later Harlow Shapley announced his conclusions about the size of our galaxy, the Milky Way. He argued that the centre of the galaxy was 50 000 light years away, much further than most astronomers had thought. Shapley was more or less correct in his estimate of the size of the Milky Way, but he was wrong in his conjecture that observed 'spiral nebulae' were within the Milky Way. As we have seen, Edwin Hubble showed that the Andromeda nebula, now called the Andromeda Galaxy, was some 800 000 light years away. This distance, which later had to be revised upwards, was so much larger than the Milky Way that it was clear that our galaxy was only one of many in the universe.

This widening of our scientific horizons was mirrored by a corresponding enlargement of the scope of science fiction. In 1928, Gernsback's *Amazing Stories* published the first instalment of a serial entitled *The Skylark of Space*, by 'Doc' Smith. The early space stories had confined themselves to the solar system: Doc Smith's 'space opera' now embraced the whole galaxy. With the launch of the magazine *Astounding Science Fiction* in 1937, edited by the legendary John Campbell, this new form of 'literature' began to attain some consistency and maturity. Probably Campbell's best known protégé was Isaac Asimov. Asimov is widely regarded as the most successful practi-

Figure 11.4 A panoramic view of the Milky Way made up from a mosaic of photographs by astronomers at the Lund Observatory in Sweden. The Andromeda Galaxy is the small elongated object in the bottom left quadrant of this picture. [Lund Observatory, Sweden.]

tioner of 'hard science fiction' – fiction that introduces and uses scientific technology in a plausible way as opposed to fantasy. Earlier in this book we quoted a description of the 'jump' through hyperspace, taken from Asimov's *Foundation* trilogy. Because relativity sets the speed of light as the maximum speed attainable, some new technology for exploring the galaxy must be invented to make feasible such galactic space operas as the *Foundation* saga. By the end of the 1950s there was a well-established convention incorporating hyperspace jumps, force shields and blasters, and this was taken forward into the television programmes and novels of the 1960s and later years. Larry Niven and Jerry Pournelle are two of the new generation of hard science fiction writers. In their now classic science fiction novel *The Mote in God's Eye*, which describes the first contact with an intelligent alien civilization, they introduce an updated version of the same theme:

> Throughout the past thousand years of history it has been traditional to regard the Alderson Drive as an unmixed blessing. Without the faster than light travel Alderson's discoveries made possible, humanity would have been trapped in the tiny prison of the Solar System when the Great Patriotic Wars destroyed the CoDominium on Earth. Instead we had already settled more than two hundred worlds. Because of the Alderson Drive we need never consider the space between the stars. Because we can shunt between stellar systems in zero time, our ships and ships' drives need cover only interplanetary distances.

Later in the novel, some plausibly sounding detail is given about hyperspace travel:

> It takes an immeasurably short time to travel between stars: but as the line of travel, or tramline, exists only along one critical path between each pair of stars (never quite a straight line but close enough to visualize it so) and the end points of the paths are far from the distortions in space caused by stars and large planetary masses, it follows that a ship spends most of its time crawling from one end point to another.
>
> Worse than that, not every pair of stars is joined by tramlines. Pathways are generated along lines of equipotential thermonuclear flux, and the presence of other stars in the geometric pattern can prevent the pathway from existing at all. Of those links that do exist, not all have been mapped. They are difficult to find.

The formula has changed little in twenty years, and the space opera is still alive and well. A cast-iron scientific basis for the faster-than-light travel and hyperspace jumps of science fiction is still lacking, although, as we shall see, Kip Thorne from Caltech appears more optimistic than most relativists. Still, in the future according to Niven and Pournelle, the Alderson Drive is not perfected till 2008... .

Asimov defines the 'Golden Age' of science fiction as being loosely the period around the 1940s and 1950s. A writer who, like Asimov, successfully made the transition from the Golden Age to the modern science fiction of today is the Englishman Arthur C. Clarke. Clarke was a technical officer in the RAF during the Second World War, and then gained a degree in mathematics and physics from London University. In 1945, he contributed an article to *Wireless World*, which demonstrated how geostationary satellites could be used to provide a global communications network. Of course, Clarke's satellites only made use of Newtonian gravity, unlike the sophisticated GPS

Figure 11.5 The cover page from an issue of Hugo Gernsback's pioneering 'pulp science fiction' magazine *Amazing Stories*. Gernsback was driven by a vision of a technologically inspired Utopia, and his magazine's mission was 'Extravagant Fiction Today . . . Cold Fact Tomorrow'. Besides reprinting classic authors like H. G. Wells and Jules Verne, *Amazing Stories* provided a platform for new writers such as 'Doc' Smith, whose famous *Skylark* series used the whole galaxy as a stage for an intergalactic 'space opera'. [*Amazing Stories* is a registered trademark of TSR Inc.]

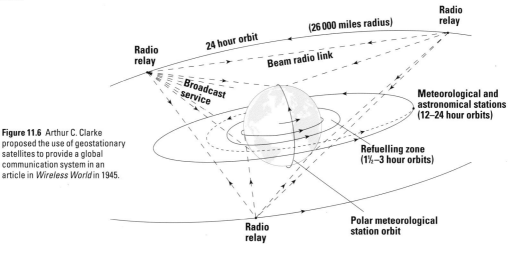

Figure 11.6 Arthur C. Clarke proposed the use of geostationary satellites to provide a global communication system in an article in *Wireless World* in 1945.

systems of today that routinely incorporate general-relativistic corrections. Clarke later became famous outside the science fiction community for his collaboration with Stanley Kubrick on the classic 1968 film *2001: A Space Odyssey*. This film is widely regarded to have set new standards for realism in its simulations of space travel.

We end this section with a look at a science fiction novel by the Frenchman Pierre Boulle, perhaps best known for his book *The Bridge on the River Kwai*, which was turned into a successful film. In 1963, he wrote a book entitled *Le Planete des Singes*, translated into English as *Monkey Planet*. Boulle's book was the inspiration for the famous 1968 film *Planet of the Apes* starring Charlton Heston. The crux of the plot is the twin paradox of relativity. In the novel, Professor Antelle explains the flight plan of his new spaceship:

> Thanks to its perfected rockets, which I had the honour of designing, this craft can move at the highest speed imaginable in the universe for a material body, that's to say the speed of light minus epsilon. I mean it can approach it [the speed of light] to within an infinitesimal degree: to within a thousandth millionth, if you care to put it that way. What you must also realize is that, while we are moving at this speed, our time diverges perceptibly from time on Earth, the divergence being greater the faster we move. At this very moment, since we started this conversation, we have lived several minutes which correspond to a passage of several months on our planet. At top speed time will almost cease to elapse for us, but of course we shall not be aware of the slightest change. A few seconds for you and me, a few heart-beats, will

coincide with a passage of several years on Earth. To reach the speed at which time almost ceases to elapse, with an acceleration acceptable to our organisms, we need about a year. A further year will be necessary to reduce our speed. Now do you understand our flight plan? Twelve months of acceleration; twelve months of reducing speed; between the two, only a few hours during which we shall cover the main part of our journey.

The key point is that hundreds of years will have elapsed on Earth while the space travellers will have aged only a few years. In the book, the astronauts travel a distance of some 300 light years to Betelgeuse – which they find populated by intelligent apes, with the roles of apes and humans reversed. Professor Antelle reverts to animal behaviour, but the journalist hero manages to convince some of the apes of his intelligence. He escapes back to Earth, only to find the apes in control. The film version cuts short this somewhat over-elaborate plot, and the initial landing is, unbeknown to the astronauts, back on Earth, centuries in the future. This is an unusually accurate use of a real effect of relativity to create an exciting and challenging story.

The present

The 1960s and 1970s saw the importance of science fiction short stories and magazines lose out to television and big budget Hollywood pictures. In the UK, *Dr Who* and the Tardis – a time machine looking on the outside like a police telephone box – pioneered the new art form in 1963. In the USA, Lucille Ball's company 'Desilu' started the *Star Trek* series in 1966. *Star Trek* brought science fiction and hyperspace to the masses: the *Starship Enterprise* and the evil empire of the Klingons are now part of modern folklore. The success of these series captured the interest of Hollywood, and the 1968 film *2001: A Space Odyssey* set new standards for special effects. In the 1970s and 1980s, George Lucas, with the *Star Wars* saga, and Stephen Spielberg, with *Close Encounters of the Third Kind* and *E.T.*, secured the success of the genre. More recently, films such as *Back to the Future* and *Bill and Ted's Excellent Adventure* have given new life to the ideas of time machines and time travel.

Let us look now at one of the most famous pieces of future technology on the *Starship Enterprise:* the matter transporter. The following anecdote illustrates how we have taken this fanciful idea to our hearts: in a recent court case in the UK, when the judge asked the defendant if he wished to say anything to the court before sentence was passed, the prisoner in the dock, using a pencil to mimic a microphone, said 'Beam me up, Scotty!' In Larry Niven's short story *Flash Crowd*, an instant riot is generated by professional rioters using 'displacement booths' to dial in and out of places in an instant. Although Niven does attempt to give some pseudo-scientific justification for teleportation in terms of an 'augmented tunnel diode effect', what defines Niven as a writer of hard science fiction is the care that he gives to explaining

Figure 11.8 Still from *Star Wars*, directed by George Lucas. [© 1996 Lucasfilm Ltd. All rights reserved.]

how such displacement booth technology could overcome the problems associated with energy and momentum conservation:

> He found that there was an inherent limitation on the augmented tunnel diode effect. Teleportation over a difference in altitude made for drastic temperature changes: a drop of seven degrees Fahrenheit for every mile upward, and vice versa, due to conservation of energy. Conservation of momentum, plus the rotation of the Earth, put a distance limit on lateral travel. A passenger flicking east would find himself kicked upward by the difference between his velocity and the Earth's. Flicking west, he would be slapped down. North and south, he would be kicked sideways.

As we will see below, it is now possible to make a much more plausible theory of teleportation in terms of 'wormhole' solutions to Einstein's field equations, but there are still worries about the conservation of energy and momentum. The *Star Trek* version of instantaneous matter transport is happy to ignore such details.

Writers of hard science fiction also pay attention to the latest theoretical and experimental discoveries. Niven's famous short story called *Neutron Star* was published in 1966, before the discovery of pulsars confirmed the existence of such objects:

Given: a burnt-out white dwarf with a mass greater than 1.44 times the mass of the Sun – Chandrasekhar's Limit, named for an Indian–American astronomer of the nineteen hundreds. In such a mass the electron pressure alone would not be able to hold the electrons back from the nuclei. Electrons would be forced against protons – to make neutrons. In one blazing explosion most of the star would change from a compressed mass of degenerate matter to a closely packed lump of neutrons: neutronium, theoretically the densest matter possible in this universe.

In the story, the puppeteers are a cowardly race of technologically advanced traders who are concerned that their spaceship business will suffer when one of their supposedly invulnerable spaceship hulls is penetrated by an unknown force that kills the occupants. This mysterious killer force turns out to be nothing other than the gigantic tidal force that would be experienced by a spaceship passing too close to a neutron star. As we have seen, Robert Forward has taken 'realism' still further in *Dragon's Egg*, his story of intelligent life on the surface of a neutron star (see also the box on p. 177).

One limitation to the mixing of time and space allowed by relativity is the preservation of causality – no moving observer can see cause and effect interchanged. The origin of this restriction may be understood using the Minkowski space-time diagrams and proper time, which we discussed in

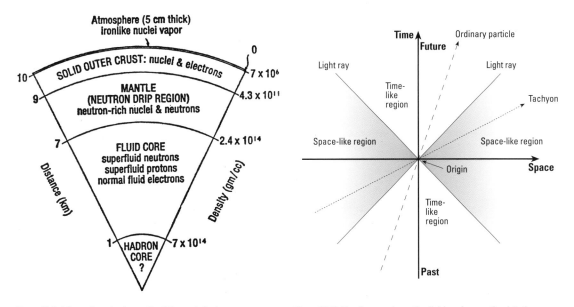

Figure 11.9 A figure from the Appendix of *Dragon's Egg* by Robert L. Forward, showing a cross-section of a segment of the neutron star. This illustrates the degree of scientific reality included by 'hard SF' writers. [From *Dragon's Egg* by Robert L. Forward (New English Library, 1981).]

Figure 11.10 The diagram shows the division of space-time into the forward (future) and backward (past) light-cones. An event at the origin of this space-time diagram – showing time and only one space dimension – can only be affected by events that have occurred in the past light-cone. For events in the past that are outside this cone, no signal travelling at less than the speed of light can reach the origin. Similarly, no signal from the event at the origin can influence events occurring outside the forward light-cone.

chapter 4. To understand which events we can influence in the future, consider a Minkowski space-time diagram with our position taken to be at the origin. No signal that we can send out or bullet that we can fire can travel faster than light. If we draw the surface mapped out by light rays sent out in all directions from our position, the light rays form the surface of a cone (Figure 11.10). Intuitively, it seems obvious that we should only be able to influence events occurring within this 'light-cone'. For such events, the square of the proper time is positive, or 'time-like'. Since the proper time is an invariant – unchanged by the relativistic 'rotations' of special relativity – causality is preserved, and cause and effect cannot get mixed up.

Of course, no one lets such a restriction stand in the way of a good story. In his novel *Timescape*, Gregory Benford has a Cambridge scientist using a beam of tachyons to send a warning back to the past. The scientist explains his apparatus to the man with the funding as follows:

> 'Well, we've got a large indium antimonide sample in there, see –'. Renfrew pointed at the encased volume between the magnetic poles. 'We hit it with high-energy ions. When the ions strike the indium they give off tachyons. It's a complex, very sensitive ion–nuclei reaction.' He glanced at Peterson. 'Tachyons are particles that travel faster than light, you know. On the other side –', he pointed around the magnets, leading Peterson to a long blue cylindrical tank that protruded ten metres away from the magnets, '– we draw out the tachyons and focus them into a beam. They have a particular energy and spin, so they resonate only with indium nuclei in a strong magnetic field.'

Figure 11.11 Still from the Walt Disney film *The Black Hole*, in which the spaceship passes through a black hole into a new, heaven-like universe. According to our present understanding of black hole physics, the ship would probably be destroyed. [© Disney Enterprises Inc.]

Such a scientific 'explanation' may well sound just as convincing as Larry Niven's description of the Chandrasekhar limit. In fact, this is just scientific

sounding mumbo-jumbo, whereas Niven is recounting current theoretical beliefs.

The success of the *Back to the Future* series of pictures is testament to the fascination exercised by the possibilities of time travel. Time paradoxes in which the time travellers can influence their own future have been a favourite theme of science fiction since H. G. Wells. *Back to the Future*, with its time-travelling De Lorean sports car, is merely the most successful screen exploration of a long tradition.

The future

In 1988, a paper with the intriguing title 'Wormholes, time machines, and the weak energy condition' was published in the prestigious scientific journal *Physical Review Letters*. The paper lay unnoticed by the general public for three months until it was discovered by a reporter from the *San Francisco Examiner*. The paper was written by a respected Caltech relativist named Kip Thorne with two of his graduate students, Mike Morris and Ulvi Yurtsever. After the *Examiner* had broken the story, headlines such as 'Physicists Invent Time Machines' became commonplace. *California* magazine even published an article on 'The man who invented time travel', with an accompanying photograph of a naked Thorne doing physics on top of Mount Palomar. Not surprisingly, Thorne went into hiding, leaving his administrative assistant to repel journalists armed only with this statement:

> Professor Thorne believes it is too early in this research effort to communicate results to the general public. When he feels he has a better understanding of whether or not time machines are forbidden by the laws of physics, he will write an article for the public, explaining.

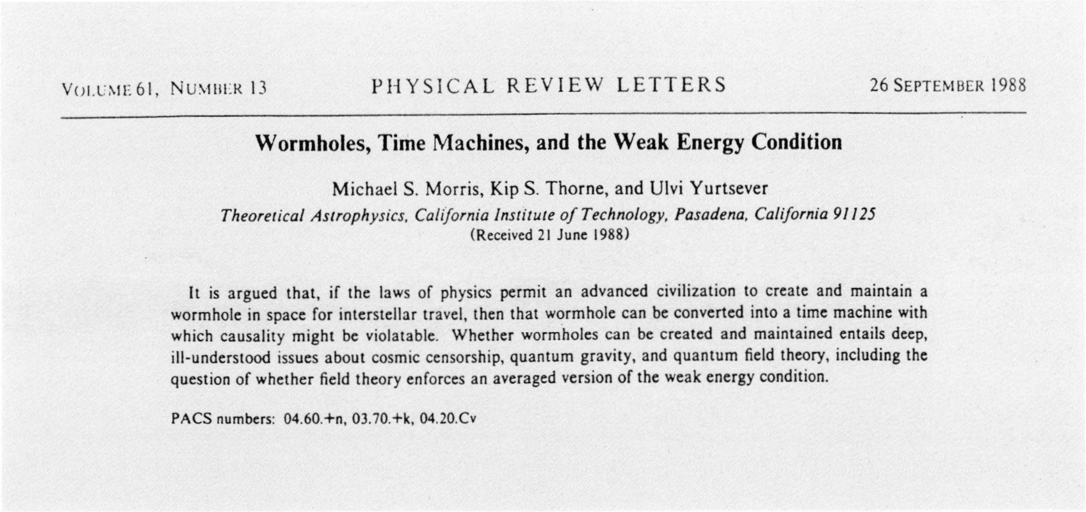

Figure 11.12 The title page of the *Physical Review Letters* article by Kip Thorne and his two students Mike Morris and Ulvi Yurtsever. Because of the press notoriety attracted by the use of the phrase 'Time Machines' in the title, Thorne now uses the phrase 'closed timelike curves' in any papers he now publishes on the subject. [Courtesy Kip Thorne.]

VOLUME 61, NUMBER 13   PHYSICAL REVIEW LETTERS   26 SEPTEMBER 1988

Wormholes, Time Machines, and the Weak Energy Condition

Michael S. Morris, Kip S. Thorne, and Ulvi Yurtsever
Theoretical Astrophysics, California Institute of Technology, Pasadena, California 91125
(Received 21 June 1988)

It is argued that, if the laws of physics permit an advanced civilization to create and maintain a wormhole in space for interstellar travel, then that wormhole can be converted into a time machine with which causality might be violatable. Whether wormholes can be created and maintained entails deep, ill-understood issues about cosmic censorship, quantum gravity, and quantum field theory, including the question of whether field theory enforces an averaged version of the weak energy condition.

PACS numbers: 04.60.+n, 03.70.+k, 04.20.Cv

Six years later, in his book *Black Holes and Time Warps: Einstein's Outrageous Legacy*, Thorne has attempted to do just that.

Thorne credits Carl Sagan with making him reflect on time machines and wormholes. Sagan was writing a book, *Contact*, in which the heroine made a hyperspace jump via a black hole. Conventional wisdom about black holes dictates that this is impossible. Thorne advised Sagan that travel through 'wormholes' in hyperspace was much more credible. A wormhole is a hypothetical connection between distant parts of the universe (Figure 11.14). They were discovered mathematically as solutions to Einstein's equations long ago, but were believed to have far too short a lifetime to be useful for space or time travel. Thorne's new ingredient is some sort of 'exotic matter' that keeps the wormhole open by exerting a gravitational *push* on the walls of the wormhole. Exotic matter lives up to its name – to some observers, at least, it must have a negative energy density. Thorne's book discusses in detail how such wormholes could be used both for hyperspace travel and for time travel. For these wormhole time machines he concludes that there are no signs of any unresolvable time paradoxes. The matter is still being debated by relativists. Stephen Hawking had gone as far as to propose the 'chronology protection' conjecture: that the laws of physics do not allow time machines – but he now appears to have recanted!

Figure 11.13 Distant star fields appear to cluster in the vicinity of a black hole, which acts like a distorting magnifying lens. Light passing near the hole is severely refracted, causing the stars to appear more luminous. Faint multiple images from every star in the universe merge to produce a glowing haze of light. [Painting by Marie Walters.]

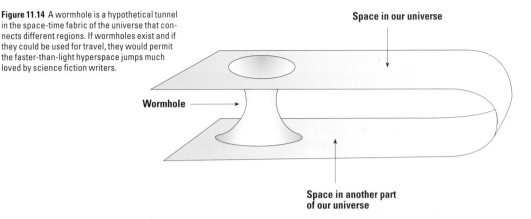

Figure 11.14 A wormhole is a hypothetical tunnel in the space-time fabric of the universe that connects different regions. If wormholes exist and if they could be used for travel, they would permit the faster-than-light hyperspace jumps much loved by science fiction writers.

Space in our universe

Wormhole

Space in another part of our universe

Given that there might be a 'respectable' basis for hyperspace jumps and time travel, it has not taken long for science fiction to use this as a basis for a novel. Robert Forward has written a book called *Timemaster*, in which the physics is firmly based on Thorne's exotic matter. In his book, Forward calls exotic matter 'negmatter' – and its anti-gravity properties are the stuff that dreams are made of:

> 'After I gave him all the facts and showed him some video segments, he conceded that maybe negative matter could exist after all. What really convinced him was the description of my injury, where the cut edges looked like a thin sliver of material that had been evaporated.'
>
> 'Why is that?' asked Randy.
>
> 'Well, as Steve explains it, according to one theory, when negative matter touches normal matter, equal amounts vanish – nothing is left, not even energy. The process is called nullification. It's like the annihilation of matter by antimatter, but in the nullification process, since the normal matter has positive rest mass and the negmatter has negative rest mass, the net rest mass is zero, so zero energy is released. That's why we didn't notice any radiation when the Silverhair and I collided.'

Forward's Silverhairs are curious negmatter plants/animals that live at either end of a wormhole and that can be persuaded to open the wormhole to allow humans to pass through. There are more surprises arising from electrically charged negmatter: an attempt to repel a positively charged negmatter ball using a positive electric charge causes it to do the opposite and come towards the charge. This property becomes the basis for both a sub-light-speed space drive and an inexhaustible source of energy. *Timemaster* presents a new view of hyperspace jumps, matter transporters and time machines, all made possible by the discovery of negmatter – a modern day version of Wells's Cavorite. The incredible journeys of Forward's hero, Randy Hunter, are accompanied by a plausible scientific subtext that challenges the reader's preconceptions:

'With the warpgate, we can cut that [the elapsed time on Earth of the twin paradox] in half,' said Randy. 'The spaceship, carrying one end of the warpgate, only has to go one way. Once there, you open up the warpgate, and Earth instantly knows what you have found.'

'It's better than that, Randy,' interjected Steve. 'The Earth shares in the astronaut's time dilation. Once the spacecraft gets up to speed, where the time-slowing factor gets large, then the spaceship can travel many light years of distance in just one year of ship proper time. For example, if the spacecraft pulls thirty gees and gets to, say, ninety-nine-point-five percent of the speed of light in a hurry, then the time-slowing factor is ten to one. The ship would cover ten light-years in one year ship time. When the astronaut uses the warpgate to return home, he will find he has been gone only one year Earth time, yet in that one year, he has opened up a warpgate to a point ten light-years distant.'

'Like travelling at ten times the speed of light,' said Randy, impressed.

These Silverhair warpgates can also be used as the basis of a time machine. Randy worries about what Thorne calls 'the martricide paradox'. His resident theorist, Steve, attempts to reassure him:

'Once any time machine starts operating, the whole future of the universe is constrained so that events happening around the time machine are consistent. For instance, the universe will arrange itself so you can't go back in time to kill yourself, no matter how hard you try.'

'Are you sure?' asked Randy.

'No,' Steve admitted. 'That's just what the theory predicts.'

'If we're not sure, then we're not going to mess with time machines,' said Randy firmly. 'I don't want to go down in history as the man who loused up the entire future history of the universe by opening up a Pandora's time box.' He turned to look around the room at everyone. 'I want you all to understand that we will only use these space warps for moving through space. Any employee found trying to use them to meddle with time will be fired and reported to the proper authorities.'

Needless to say, Randy is forced to change his mind.

Kip Thorne's speculations about exotic matter and wormholes are indeed still just speculations. In his book he states:

If time machines are, in fact, allowed by the laws of physics (and, as will become clear at the end of the chapter, I doubt that they are), then they are probably much farther beyond the human race's present technological capabilities than space travel was beyond the capabilities of cavemen.

The problem has to do with our lack of an understanding of how to merge quantum mechanics and gravity to produce a consistent theory of quantum gravity. It is entirely possible that wormholes, and even black hole singularities, may be destroyed by the effects of quantum fluctuations. On the other hand, there is always hope that quantum gravity will lead to dramatic new insights about the structure of our universe. Until we know the answer, there will always be healthy speculation – both within the 'serious' scientific literature, and without, in the realm of science fiction.

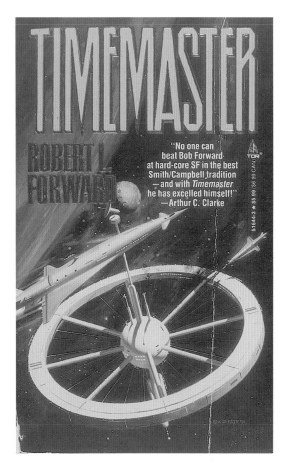

Figure 11.15 The cover picture of *Timemaster*, the novel by Robert L. Forward, that uses wormholes and a modern variant of Cavorite for faster-than-light travel and for time travel. [Timemaster, Robert L. Forward (Tom Doherty Associates, Inc., New York, 1992). Cover art by Vincent Di Fate.]

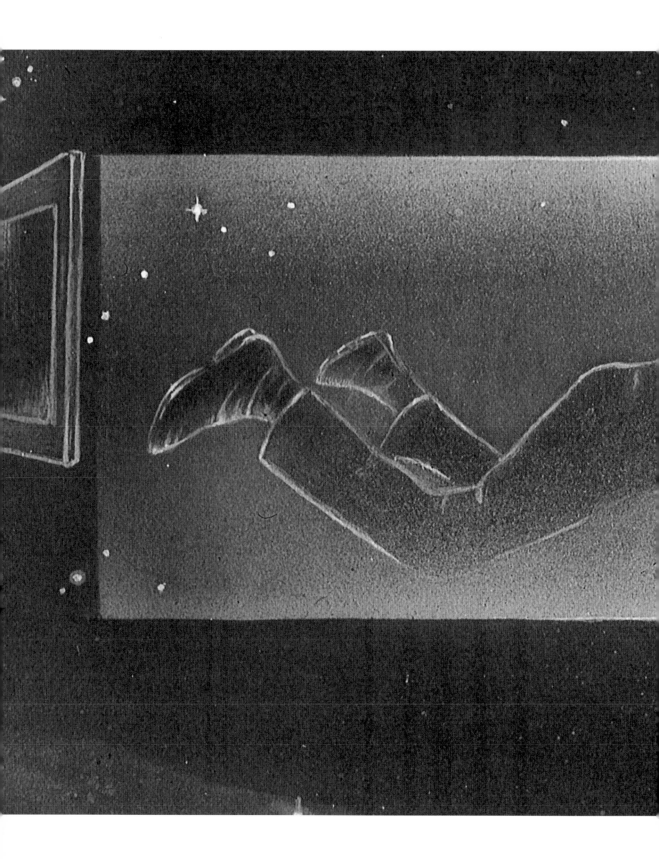

Appendix **Some mathematical details and derivations**

Time dilation

In chapter 3, we promised more details on the derivation of relativistic 'time dilation'. We considered a thought experiment with a simple clock and two 'observers' – one at rest relative to the clock and one in motion. We shall now see that the stationary observer sees the moving clock running slow.

Our 'clock' consists of a box with a mirror at either end. A light pulse bounces back and forth between the two mirrors, and the time taken for one round trip is taken to be one 'tick' of the clock. For an observer at rest relative to this clock, all is fine. Consider how things look for an observer moving in a direction at right angles to the length of the clock (Figure A1). She will see the light pulse start out, and she will see the clock with its two mirrors move away from her with a constant speed. According to her, the light must travel further than just twice the distance between the mirrors. The argument is very similar to that used in the Michelson and Morley experiment. If the distance between the mirrors is denoted by L, and the speed at which the clock moves away from our observer is written as v, we can relate the times measured by the two observers: T_R for the observer at rest with respect to the mirrors, and T_M for the observer in motion. We have

$$T_R = 2L/c$$

and

$$T_M = 2[L^2 + (vT_M/2)^2]^{1/2}/c$$

where we have denoted the velocity of light by its usual symbol, c. Note that we have assumed that Einstein is correct: both observers measure the same value for the velocity of light. We must now manipulate these equations so that we can eliminate L and relate T_R to T_M. Square both equations and rearrange the second to collect together all terms with T_M:

$$(cT_R)^2 = 4L^2$$

$$(T_M)^2(c^2 - v^2) = 4L^2$$

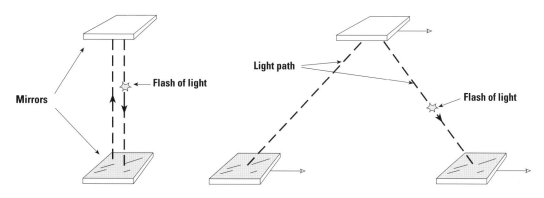

Mirrors

← **Flash of light**

Light path

Flash of light

Figure A1 A light clock: one unit of time corresponds to a pulse of light travelling from one mirror to the other and back, as shown by the dashed line. When in motion the light path is the longer triangular dashed line shown. Since the light now travels a longer distance when the clock is in motion, the moving clock runs slower than an otherwise identical stationary clock.

Equating these expressions for $4L^2$ and doing a little rearrangement, we obtain the famous result

$$T_M = T_R / [1 - v^2/c^2)]^{1/2}$$

This is time dilation: *moving clocks run slow.*

Velocity addition

We discussed the addition of velocities according to special relativity in chapter 4. Relativity predicts that the simple addition law of Newton is incorrect – although it is obviously a very good approximation for speeds much less than the speed of light. There are many ways to derive Einstein's formula, and we shall follow a method due to David Mermin, published in his thought-provoking collection of essays entitled *Boojums All The Way Through*.

Mermin's derivation of Einstein's velocity addition formula is beautiful in its simplicity and elegance. His thought experiment consists of a long train moving parallel to its own length with speed v (Figure A2). We shall describe all events from the viewpoint of an observer standing at rest by the track. An observer on the train starts a race from the rear of the train to the front, between a light photon, which always travels with speed c, and a bullet, which travels at a speed w, where w is necessarily less than the speed of light. The photon obviously returns along its path and meets up with the speeding bullet on its way to the front of the train a fraction f of the length of the train, as measured from the front. Both the observer on the train and the observer on the ground must agree on this value of f. We can calculate this fraction from the following three observations.

1. The total distance the bullet has moved from the start of the race to its encounter with the photon must be equal to the distance the photon has moved in going from the rear of the train to the front, minus the distance the photon moved in going back towards the rear. If we write T_1 for the time

taken for the photon to reach the front of the train, and T_2 for the time taken from the reflection of the photon till it meets the bullet, we can express this fact as follows:

$$w(T_1+T_2)=c(T_1-T_2) \tag{1}$$

2. The distance the photon has moved in going from the rear of the train to the front is the length of the train plus the distance the train has moved during this time T_1. If L is the length of the train, we can express this as:

$$cT_1=L+vT_1 \tag{2}$$

3. The distance the photon has moved in going from the front of the train back to meet the bullet is the length of the train from the front to the encounter point, f times L, minus the distance the train travelled in the time T_2. As an equation, this fact can be written:

$$cT_2=fL-vT_2 \tag{3}$$

Equation (1) can be rearranged to give an expression for the ratio of T_2 to T_1:

$$T_2/T_1=(c-w)/(c+w)$$

Equations (2) and (3) can be used to eliminate the length L. After some manipulation, this yields another expression for the ratio T_2/T_1:

$$T_2/T_1=f(c-v)/(c+v)$$

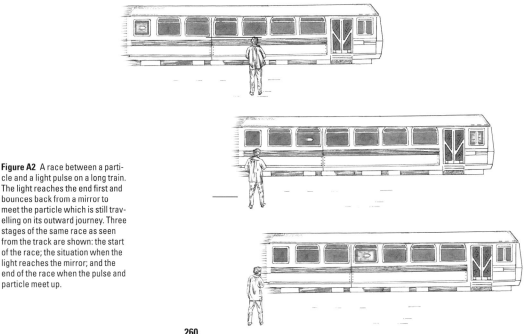

Figure A2 A race between a particle and a light pulse on a long train. The light reaches the end first and bounces back from a mirror to meet the particle which is still travelling on its outward journey. Three stages of the same race as seen from the track are shown: the start of the race; the situation when the light reaches the mirror; and the end of the race when the pulse and particle meet up.

Equating these two expressions yields the following formula for the fraction f:

$$f = \frac{(c+v)(c-w)}{(c-v)(c+w)}$$

This is the key result. Nowhere in the derivation was v, the speed of the train, required to be non-zero. We can therefore use the same formula to calculate f for an observer at rest with respect to the train. In this frame, we denote the speed of the bullet according to this observer as u: the speed of light and the fraction are the same for any observer. In terms of the speed u, the fraction f is given by

$$f = \frac{(c-u)}{(c+u)}$$

Equating these two expressions for f gives the result we want: the relationship between the speed of the bullet as seen by an observer at rest on the train, u, and the speed, w, seen by an observer moving with velocity v with respect to the train. We have:

$$\frac{(c-u)}{(c+u)} = \frac{(c+v)(c-w)}{(c-v)(c+w)}$$

Solving for w gives the desired result:

$$w = \frac{(u+v)}{1+(uv/c^2)}$$

If the velocity of light were infinite, the second term in the denominator would vanish, and the result reduces to the usual Newtonian answer:

$$w = u+v$$

For velocities that are small compared with the velocity of light, the formula also shows that Newtonian addition of velocities will be a very good approximation.

The relativistic mass increase

Probably the most famous formula in physics is Einstein's relation connecting mass and energy. Einstein arrived at his formula via a thought experiment inspired by his observation that the mass of a moving particle depends on its velocity. In chapter 5 we discussed another thought experiment involving two basketball players and a fast-moving train. Here we give a few more details of the mathematics that led to Einstein's conclusion.

We recall that we have two highly skilled basketball players and we ask them to perform two experiments (Figure A3). In the first experiment, one player stands on a stationary train by an open window and the other stands opposite him, stationary by the side of the track. Each player has an identical basketball, and they throw their balls towards each other. They do this at exactly the same moment, with exactly the same speeds, so that each trajectory is a mirror image of the other. Halfway between the players, the balls collide and bounce back along exactly the same paths into the hands of the two players. There is nothing surprising about this – it is exactly what we expect from the 'conservation of momentum'. Momentum is mass times velocity, and each ball starts with a momentum $m_0 u_0$, where m_0 is the mass of the ball and u_0 is the initial velocity. After the collision, each ball's momentum is reversed so the momentum change of each ball is $2m_0 u_0$. The momentum change of each ball is equal, and we may therefore write conservation of momentum as follows:

$$2m_0 u_0 = 2m_0 u_0$$

(momentum change for ball 1) = (momentum change for ball 2)

We have ignored real-life complications due to gravity and the fact that the throwers have to throw the balls slightly upwards as well as outwards. Such complications do not affect the essential argument.

The second experiment would really require a lot of skill from the two players, but it is possible in principle, which is the essence of a thought experiment. The players have to repeat their trick when the train is moving at high speed past the player standing by the track. According to Newton's laws,

Figure A3 The top picture shows a basketball player on a stationary train and a colleague on the platform, each throwing a basket ball, which subsequently collide and bounce back to be caught by the players. The bottom illustration shows the more difficult experiment performed from the window of a rapidly moving train. The arrows show the trajectory of the balls. Notice that the ball thrown from the moving train continues to move rapidly in tandem with the train. However, from the point of view of the stationary player his colleague seems to throw the ball more slowly because of 'time dilation'. (Relativistic shortening of the train has been ignored.)

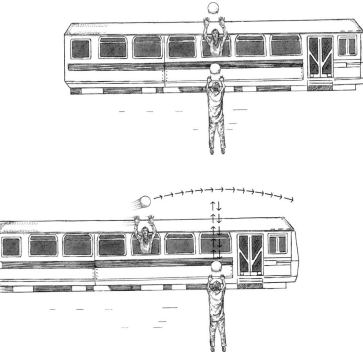

if the players time their throws just right to allow for the sideways motion of the train, conservation of momentum in the direction at right angles to the motion of the train will be just as before. What goes wrong for trains moving with relativistic high speeds? Consider the experiment as viewed by the player standing by the track. He sees the player on the train throw the ball out too slowly. The fact that he has a slower speed than before can be seen either by a generalization of the velocity formula derived above or by a time dilation argument. Whichever way we perform the calculations, we find that he sees the velocity reduced by the time dilation factor derived above:

$$\frac{1}{[1-(v^2/c^2)]^{1/2}}$$

where v is the speed of the train. Thus the velocity seen by the player on the track is not u_0 but u_1:

$$u_1 = u_0[1-(v^2/c^2)]^{1/2}$$

Although the player by the track thinks that the ball has been thrown from the train too slowly for the experiment, he is very surprised to find that things have been judged just right, and his ball returns to him as before.

We could also analyse the problem from the point of view of the player on the train. He throws his ball out of the window with just the same speed as before. He is dismayed to see that his friend by the track seems to have misjudged things and thrown the ball towards the train with too small a speed. Again the player on the train is surprised to find that the experiment works. The inconsistency appears the same from both points of view.

What is the way out? It appears that conservation of momentum is in trouble. The player by the track still sees a momentum change of $2m_0u_0$ for his ball. How can this be balanced by the momentum change of the ball of the player on the train if the velocity of that ball is reduced from u_0 to u_1? The only way relativistic momentum conservation can be restored is for the mass of the ball of the moving player on the train to increase by exactly the same factor! Momentum conservation now reads:

$$2m_0u_0 = 2\{m_0/[1-(v^2/c^2)]^{1/2}\}\{u_0[1-(v^2/c^2)]^{1/2}\}$$

We have arrived at the conclusion that the mass of a ball increases with velocity according to the formula

$$m_v = \frac{m_0}{[1-(v^2/c^2)]^{1/2}}$$

where m_v is the mass of the ball moving with velocity v.

When an object is accelerated to very high speeds, most of the energy used to try to accelerate the object must be used to increase the mass of the moving particle. It was from consideration of this formula that Einstein was led to his famous mass–energy relation.

Chronology

Our story of Einstein's discovery of both special and general relativity covers many aspects of physics and astronomy. To help the reader locate the position in time of the different events and important ideas along the way, we offer this brief (and incomplete) 'chronology' as a guide.

Astronomy

Aristotle (384–322 BC)   the Earth is the centre of the Universe
Nicolaus Copernicus (1473–1543)   the Sun is the centre of the solar system (published 1543)
Tycho Brahe (1546–1601)   accurate measurements of stars and planets

Gravity

Johannes Kepler (1571–1630)   laws of planetary motion (1609 and 1619)
Galileo Galilei (1564–1642)   law of free fall (1638)
Isaac Newton (1642–1727)   theory of gravity (published 1687)

Mechanics

Galileo Galilei (1564–1642)   principle of 'Galilean' relativity (1632)
Isaac Newton (1642–1727)   Newton's laws of motion (published 1687)
Ernst Mach (1836–1916)   critique of absolute space and time (1863)

Light

Christian Huygens (1629–1695)   wave theory of light (1678)
Ole Roemer (1644–1710)   first measurement of the speed of light (1676)
Albert Michelson (1852–1931)   accurate measurement of speed of light

Electricity and magnetism

Michael Faraday (1791–1867)   discovery of electromagnetic induction (1831)
James Clerk Maxwell (1831–1879)   Maxwell's equations and light as an electromagnetic wave (1864)
Heinrich Hertz (1857–1894)   discovery of radio waves (1888)

Special relativity

Albert Michelson (1852–1931)
    and Edward Morley (1838–1923)   aether drift experiment (1881 and 1887)

George Francis Fitzgerald (1851–1901)   Lorentz–Fitzgerald contraction hypothesis (1887)
    and Hendrik Antoon Lorentz (1853–1928)

Joseph Larmor (1857–1942)   local and general time (1898)

Jules Henri Poincaré (1854–1912)   relativistic velocity addition (1905)

Albert Einstein (1879–1955)   theory of special relativity (1905) and mass–energy equivalence (1906)

Herman Minkowski (1864–1909)   four-dimensional space-time (1908)

Energy

Georg Stahl (1660–1734)   phlogiston theory

Joseph Priestley (1733–1804)   discovery of oxygen (1774)

Antoine Laurent Lavoisier (1743–1794)   conservation of mass

Benjamin Thomson (Count Rumford)   heat as internal atomic motion
    (1753–1814)

Julius Robert Mayer (1814–1878)   conservation of energy

James Prescott Joule (1818–1889)   mechanical equivalence of heat

William Thomson (Lord Kelvin) (1842–1907)   first law of thermodynamics

Sadi Carnot (1796–1832)   second law of thermodynamics (1824)

Rudolf Clausius (1822–1888)   concept of entropy

Atoms

Daniel Bernoulli (1700–1782)   kinetic theory of gases (1738)

John Dalton (1766–1844)   atoms and molecules in chemistry

Ludwig Boltzmann (1844–1906)   atomic explanation of entropy as disorder (1877)

Josiah Willard Gibbs (1839–1903)   foundations of statistical mechanics

Albert Einstein (1879–1955)   atomic picture of Brownian motion (1905)

Jean Baptiste Perrin (1870–1942)   experimental confirmation of theory of Brownian motion (1908)

Nuclear physics

Wilhelm Conrad Roentgen (1845–1923)   X-rays (1895)

Henri Becquerel (1852–1908)   radioactivity (1896)

Marie Curie (1867–1934)   radioactive decay and half-lives of elements (1898)
    and Pierre Curie (1859–1906)

Ernest Rutherford (1871–1937)   alpha and beta rays (1898); radioactive transformations (1903); the nucleus (1911);
                                                     nuclear reactions (1919)

Paul-Ulrich Villard (1860–1934)   gamma rays (1900)

Francis Aston (1877–1945)   mass spectrometer and isotopes (1916)

Ernest Lawrence (1901–1958)   cyclotron (1931)

John Cockcroft (1897–1967) and Ernest Walton (*b*1903)	first artificially induced nuclear reaction (1932)
Leo Szilard (1898–1964)	idea of nuclear chain reaction (1934)
Frederic Joliot (1900–1958) and Irene Joliot-Curie (1897–1956)	artifically induced radioactivity (1934)
Enrico Fermi (1901–1954)	radioactive isotopes using neutrons (1934)
Ida Noddack (*b*1896)	suggestion of nuclear fission (1935)
Otto Hahn (1879–1968) and Fritz Strassman (1902–1980)	discovery of nuclear fission (1938)
Lise Meitner (1878–1968) and Otto Frisch (1904–1979)	interpretation of nuclear fission (1939)
Rudolf Peierls (1907–1995)	Frisch–Peierls memorandum showing nuclear bombs are feasible (1940)

Particle physics

Joseph John Thomson (1856–1940)	discovery of the electron (1897)
Ernest Rutherford (1871–1937)	discovery of the proton (1919)
James Chadwick (1891–1974)	discovery of the neutron (1932)
Paul Dirac (1902–1984)	existence of anti-matter predicted (1931)
Carl Anderson (1905–1991)	discovery of the positron or anti-electron (1932)
Owen Chamberlain (*b*1920), Emilio Segrè (1905–1989) and others	discovery of the anti-proton (1955)

Quantum theory

Niels Bohr (1885–1962)	'solar system' model of the atom (1913)
Arnold Sommerfeld (1868–1951)	atomic 'fine structure' (1915)
Werner Heisenberg (1901–1976)	matrix mechanics (1925); uncertainty principle (1927)
Erwin Schroedinger (1887–1961)	Schroedinger equation for wave mechanics (1925)
Max Born (1882–1970)	probability interpretation of wave function (1926)
George Uhlenbeck (1900–1988) and Samuel Goudsmit (1902–1978)	electron spin (1925)
Paul Dirac (1902–1984)	Dirac equation (1928)

Non-Euclidean geometry

Euclid (*c*300 BC)	elements of geometry
Nicolai Ivanovitch Lobachevsky (1793–1856), Janos Bolyai (1802–1860) and Karl Friedrich Gauss (1777–1855)	non-Euclidean geometry (*c*1826)
Georg Friedrich Riemann (1826–1866)	many-dimensional geometries (1854)

General relativity

Johann Von Soldner (1776–1833)	bending of light by the Sun according to Newtonian gravity (1801)
Urbain Jean Joseph LeVerrier (1811–1877)	anomalous perihelion precession of orbit of Mercury (1859)
Roland Eotvos (1848–1919)	equivalence of inertial and gravitational mass (1889 and 1908)
Albert Einstein (1879–1955)	equivalence principle (1907); general theory of relativity (1916)
Arthur Stanley Eddington (1882–1944)	eclipse expedition to verify bending of light prediction (1919)
Robert Dicke (*b*1916)	Brans–Dicke theory of gravity (1960)
Robert Pound and Glen Rebka	measurement of gravitational red-shift (1960 and 1965)

Irwin Shapiro (b1929)	time delay of light passing near the Sun as new test of general relativity (1967)
Kenneth Nordtvedt	'Nordtvedt effect' and Brans–Dicke theory (1967)

Radio astronomy

Karl Jansky (1905–1950)	radio waves from the Milky Way (1931)
Grote Reber (b1911)	discrete radio sources (1937)
James Stanley Hey (b1909)	Sun as a radio source (1942); first extra-galactic radio source (1946)

Cosmology

Albert Einstein (1879–1955)	cosmological constant (1917)
Aleksandr Friedmann (1888–1925)	general relativistic models of an expanding universe (1922)
Edwin Hubble (1889–1953)	Andromeda Galaxy (1925); Hubble's law (1929)
Georges Lemaître (1894–1966)	Big Bang theory (1933 and 1946)
George Gamow (1904–1968), Ralph Alpher (b1921) and Robert Herman	hot Big Bang model (c 1946)
Hermann Bondi (b1919), Thomas Gold (b1920) and Fred Hoyle (b1915)	steady state model (1948)
Arno Penzias (b1933) and Robert Wilson (b1936)	cosmic microwave background radiation (1965)

Black holes

John Michell (c1724–1793) and Pierre Simon de Laplace (1749–1827)	idea of Newtonian black hole (1783 and 1795)
Karl Schwarzschild (1876–1916)	non-spinning Einstein black hole (1916)
Robert Oppenheimer (1904–1967) and Hartland Snyder (1913–1962)	imploding star can form black hole (1939)
John Archibald Wheeler (b1911)	coins name 'black hole' (1967)

Unified Field Theory

Emmy Noether (1882–1935)	relationship between symmetries and conservation laws (1918)
Theodor Kaluza (1885–1954) and Oskar Klein (1894–1977)	unified theory of gravity and electromagnetism in five-dimensional space (1921 and 1926)
Sheldon Glashow (b1932), Abdus Salam (b1926) and Steven Weinberg (b1933)	unified electro-weak model (1967)

Glossary

absolute time
the idea suggested by Isaac Newton that time is independent of anything else (such as motion or gravity) and is the same throughout the universe.

absorption spectrum
dark lines in the spectrum of some object, such as a star, due to the absorption of light by material along its path. For example, the material could be in the outer layers of the emitting star or a gas cloud closer to the observer. Like emission lines, the pattern of these lines reveals the chemical composition of the material. *See* emission spectrum.

alpha particle
a helium nucleus, which consists of two *protons* and two *neutrons*. In alpha radioactive decay an unstable *nucleus* emits a rapidly moving alpha particle.

anode
positively charged conductor through which an electric current enters a gas, vacuum or conducting substance such as an electrolyte. *See* cathode.

anti-matter
anti-matter is made of anti-particles, which have the same mass as the particle concerned, but opposite 'charge-like' properties. When a particle and its anti-particle meet, they can annihilate to become energy. There is an anti-neutron, despite the zero electric charge of the *neutron*, because neutrons have a property corresponding to the conservation of the total number of baryons in the universe. Anti-neutrons have the opposite value of this 'baryonic charge'.

arrow of time
the idea that time has a sense of direction – it 'flows' one way. Because *entropy* is always increasing, this can give a direction to the arrow of time.

atomic mass unit
one-twelfth of the mass of the neutral carbon 12 atom. In these units, a *proton* has an atomic mass of 1.00728.

beta particle beta decay is a radioactive process in which a *neutron* (or *proton*) is changed to a proton (or neutron) emitting an *electron* (or anti-electron). The emitted electron is called a beta particle. Beta radioactivity proceeds by the *weak interaction*.

binding energy the energy needed to separate completely a nucleus into its constituent *protons* and *neutrons*. The binding energy per *nucleon* indicates how tightly the nucleus is bound.

black body the name given to an ideal radiator, which also absorbs all the radiation falling on it. It emits a continuous spectrum of radiation, which depends only on the temperature. The total power radiated increases very rapidly with rising temperature and the peak intensity of radiation moves towards the shorter wavelength end of the *electromagnetic spectrum of radiation* as the temperature increases.

black hole an object where gravity dominates all other forces and produces a collapse to a singularity, where the known laws of physics break down. Once light or anything else enters a critical region about the singularity, at a distance called the *Schwarzschild radius*, it can never escape. This property gives rise to the name 'black' hole.

Brownian motion small erratic and ceaseless motion of particles suspended in a liquid or gas. First observed by Robert Brown (1773–1858) using pollen in water.

caloric the false concept of heat as a substance.

cathode negatively charged conductor through which an electric current leaves a gas, vacuum or conducting substance such as an electrolyte. *See* anode.

causality the idea that every event must have resulted from other earlier events or causes. All physical causes travel at the speed of light or less. This idea is called into question by *quantum mechanics*.

Chandrasekhar limit the maximum mass of a *white dwarf* star.

classical theories physics before quantum theory. The key areas of classical physics are Newtonian mechanics, electromagnetic theory and *thermodynamics. Special* and *general relativity* are sometimes regarded as a completion of classical physics, unifying its different aspects, particularly mechanics and *electromagnetism*.

cosmology the study of the universe as a whole.

critical mass the minimum mass of a given fissionable substance, such as uranium 235, able to sustain a nuclear chain reaction.

cross-section the effective area for the collision of two particles to produce some specific products. In an elastic interaction the two particles emerge from their encounter deflected, but unscathed. The elastic cross-section for an *alpha particle* in collision with an atom will be much smaller than the atom and quite close to the area presented by the nucleus.

Doppler effect the change in the wavelength of waves arriving at a receiver when the source and receiver are in relative motion. The Doppler shift causes an increase in the wavelength when the source and receiver are moving apart and, conversely, a decrease in wavelength when they approach.

Einstein's mass–energy relation, $E=mc^2$ this formula means that matter is frozen energy, concentrated in a small region. Small changes in the mass of objects, such as atomic nuclei, can therefore release large amounts of energy.

electromagnetic spectrum of radiation waves of electric and magnetic fields. The longest wavelength waves are called radio waves and successively shorter wavelengths include infrared, visible light, ultra-violet, X-rays and gamma rays. All components of the electromagnetic spectrum travel at the speed of light in a vacuum.

electromagnetism one of the fundamental forces of nature which occurs when charged particles inter-act. The electrical attraction between *electrons* and *protons* which holds atoms together is an example of the electromagnetic force.

electron a negatively charged elementary particle with no strong interaction (with the nucleus). Electrons are constituents of all atoms. They surround the central *nucleus* and give the atom its size, strength and chemical properties. Electrons are very light compared with the nucleus.

electron spin the quantum property of *electrons* and many other particles corresponding to the clas-sical idea of rotation. This property is quantized, and electrons have the lowest possi-ble non-zero spin value.

electroweak theory theory unifying *electromagnetism* and the *weak nuclear force*. The relationship between the two forces is hidden at normal energies, but at high energies, such as prevailed in the early universe, the two forces function as a single electroweak force.

emission spectrum the spectrum of radiation produced by excited atoms or molecules. The excitation can be produced in a variety of ways including burning the substance or causing the vapour to glow in an electric discharge. The bright lines in the spectrum are character-istic of the atom or molecule concerned and act as an atomic 'fingerprint'. *See* absorp-tion spectrum.

entropy a measure of the 'disorder' in a large system of particles. Originally introduced in con-nection with the second law of *thermodynamics*.

equivalence principle the idea that the physics in an accelerated laboratory is equivalent to that in a uniform gravitational field. In particular, in a freely faling laboratory, the laws of physics, including *special relativity*, are the same as in a gravity-free region.

event horizon the 'surface' of a *black hole*, which is the boundary beyond which nothing can emerge.

fine structure when a spectral line or band is viewed at high resolution, fine structure is observed. For example, the yellow line of the sodium spectrum is actually two closely spaced lines when viewed in detail. This structure is caused by the *electron spin*. 'Hyperfine structure' corresponds to even more closely spaced spectral lines caused by the inter-action of the spin of the nucleus with that of the electron, or different *isotopes* within the sample.

fluorescence light produced by some method other than heating (luminescence), but which ceases as soon as the source of excitation stops. For example, a domestic fluorescent light consists of a gas discharge tube whose interior is coated with a thin layer of phosphor. The impacts of electrons from the electric discharge cause the phosphor to emit light or *fluoresce*. *See* phosphorescence.

gamma rays

light of very short wavelength whose quanta – *photons* – have a very high energy. Gamma rays are produced by some radioactive decays.

gauge invariance

this was originally introduced for certain theories that are size or scale independent. When quantum theory was introduced, the idea of gauge invariance became a more abstract symmetry, which, if satisfied, implied that the corresponding force quanta are massless. Electromagnetic theory is gauge invariant and its quanta – *photons* – are massless.

gauge theory

Einstein's *general relativity* is a gauge theory, which means that the force of *gravity* can be locally eliminated by considering the region from the point of view of a freely falling observer. This idea is the essence of the *equivalence principle*. *Electromagnetism* can be considered in a similar way, but the local transformation needed to locally eliminate the electromagnetic field is more abstract. A similar approach applies to the theories of the *weak force* and *quantum chromodynamics*.

gedanken experiment

gedanken is a German word which means 'thought'. Hence, a gedanken experiment is carried out in the imagination rather than in practice. Einstein was responsible for many 'thought experiments', such as the one we have called 'Einstein's mirror'.

general relativity

Einstein's theory of *gravity* as an expression of the curvature of *space-time*.

geodesics

the straightest possible lines in curved space or curved *space-time*. On a sphere the geodesics are great circles.

Grand Unified Theory (GUT)

theory which suggests a link between the *electroweak* and *strong forces*. These theories work in analogy with the successful *electroweak theory*. However, the relationship between the two forces remains hidden until extremely high energies, beyond any conceivable accelerator experiments.

gravity

Newton described gravity as the mutual attraction of all massive objects, acting instantaneously across space. *General relativity* describes gravity as a consequence of the curvature of *space-time* caused by massive objects. In this field theory, the effects of gravity are conveyed at the speed of light.

half-life

the time taken for half the atoms in a specific radioactive material to disintegrate.

heavy water

a substance which is chemically the same as water, but in which all the hydrogen atoms are replaced with a heavier *isotope*, deuterium.

Hubble constant (H)

the constant of proportionality in *Hubble's law*. Uncertainties concerning the distance of galaxies results in a similar doubt about the value of the Hubble constant. This results in a value somewhere within the region of about 10–30 kilometres/second/million light years.

Hubble's law

the recession velocity (v) of a distant galaxy increases uniformly with distance (d): $v=Hd$, where H is called the Hubble constant.

inertial mass

a measure of the tendency of a body to resist changes in its motion.

interferometer

apparatus in which a beam of light or other radiation is split and travels along two paths before being reunited. In the region where the beams recombine, an interference pattern results. This is because in some places the two waves will be oscillating in tandem, producing more intense radiation, while in other regions the waves will be out of step and will cancel out (interfere), resulting in a less intense region. The Michelson and Morley apparatus was an interferometer.

isotope

nuclei with the same number of *protons* but different numbers of *neutrons* are the isotopes of a given element.

Larmor time dilation

the increase in the period of a clock in motion through the aether. A similar time dilation occurs in relativity, but without reference to the *luminiferous aether*.

light speed

in *special relativity*, the speed of light in a vacuum is a universal constant for all uniformly moving observers. It is about 300 000 kilometres/second.

lines of force

a representation of the direction of a field of force by lines. The density of the lines indicates the intensity of the field. This picture is especially useful when these lines are not created or destroyed, such as static electric or magnetic fields.

Lorentz–Fitzgerald contraction

a contraction of a body along its direction of motion, postulated to occur when travelling through the *luminiferous aether*. The same contraction of objects in motion occurs in *relativity*, but the explanation does not involve the hypothetical *aether*.

Lorentz transformations

the special relativistic relationship between the *space-time* coordinates of two observers in relative motion.

luminiferous aether

aether was a hypothetical medium, the undulations of which were interpreted as light. For this reason it was often referred to as luminiferous aether.

meson

any strongly interacting boson. All mesons are unstable and consist of a bound state of a *quark* and an antiquark.

microwave background radiation

black body radiation at a temperature of about 2.7 kelvin, coming from all directions in the sky. Its properties coincide with the red-shifted relic of the Big Bang fireball when it was less than one million years old.

mini black holes

a hypothetical *black hole* with a mass less than about one hundred million tons and a *Schwarzschild radius* smaller than an atom. These objects create such intense local gravitational tides that quantum effects cause them to radiate and finally explode.

neutron

electrically neutral constituent of the nucleus with almost the same mass as the *proton*.

neutron star

a star composed predominantly of neutrons and with a diameter of about 10 miles and mass comparable with the Sun. Neutron stars are believed to be formed as a consequence of *supernova* explosions.

nuclear fission

the division of a large *nucleus* into two roughly equal moderate sized nuclei with possible small fragments in addition. Nuclear fission can be induced in certain nuclei by firing neutrons into them; this forms the basis for nuclear reactors.

nuclear fusion	any nuclear reaction in which two or more nuclei come together to form a nucleus with more neutrons plus protons than any of the ingredients. Fusion reactions of hydrogen to make helium are the energy source for the Sun.
nucleon	a constituent of the *nucleus* – a *proton* or a *neutron*.
nucleus	dense core of an atom made of *protons* and *neutrons* held together by the *strong nuclear force*.
Pauli Exclusion Principle	roughly, this is the idea that 'matter particles' (strictly fermions such as *electrons*) avoid each other, or, more precisely, must have different quantum numbers. This principle is the key to understanding such diverse concepts as the *Periodic Table* of the elements and *neutron stars*.
Periodic Table	tabulation of the different kinds of atoms (elements) in terms of increasing numbers of *protons* in the *nucleus*. Atoms with similar chemical and physical properties occur regularly and are placed in columns.
perpetual motion machine	a machine that would keep going indefinitely. If such a machine could perform useful work, it would violate the law of conservation of energy, and its impossibility is equivalent to the first law of *thermodynamics*. It is also impossible to construct a machine which converts heat entirely into work. This is a perpetual motion machine of the second kind, and the second law of *thermodynamics* can be stated in terms of the prohibition of such devices.
phlogiston	the eighteenth century idea that combustion involved the loss of a substance – phlogiston.
phosphor	any substance that emits light when subject to some process (for example collision by electrons) other than heat. Such substances will display *fluorescence* or *phosphorescence*.
phosphorescence	light emitted from a substance after the source of excitation has been removed. *See* fluorescence.
photo-electric effect	the emission of *electrons* from a material, caused by light or other electromagnetic radiation incident upon it. The effect was explained by Einstein in terms of light considered as a stream of corpuscles called *photons*.
photon	the quantum particle associated with light waves or, more generally, electromagnetic radiation.
pion (pi-meson)	the lightest *meson*. It comes in three varieties distinguished by an electric charge equal to 1, 0 and −1 times that of the *proton*.
Planck's constant	the fundamental constant of *quantum mechanics*.
planetary nebula	an expanding cloud of gas that is believed to be ejected from red giants late in their evolution. A burnt out stellar core remains at the centre of the cloud, and ultraviolet radiation from it ionizes the gas. It receives its name because the cloud of gas is usually symmetrical and so gives the appearance of a disc, like a planet.
positron	positively charged anti-particle of the *electron*.

proton   the positively charged constituent of the *nucleus*.

pulsar   rapidly varying source of radio waves, presumed to be due to a rotating *neutron* star.

quantum chromodynamics   the quantum theory of the interaction of *quarks*, which carry a property similar to electric charge called 'colour' (although it has nothing to do with the familiar concept of colour).

quantum electrodynamics   the quantum theory of the electromagnetic interaction.

quantum mechanics   the fundamental physical theory describing the behaviour of all objects, but whose effects are normally only apparent in the realm of the very small (atoms, molecules, electrons, photons, etc.). Quantum effects are characterized by a quantity called *Planck's constant* and include such things as discrete energy levels in atoms (quantization) and intrinsic uncertainty about the value of certain dual quantities such as position and velocity.

quarks   the fundamental particles which are thought to form the basic building blocks of strongly interacting matter. *Protons* are made of three quarks.

quasars   a star-like object with a large *red-shift*. If the red-shift is interpreted as due to the Hubble expansion of the universe, then quasars radiate about 100 times the energy of a conventional galaxy from a central region – or active galactic nucleus – not much larger than the solar system! Massive *black holes* are involved in most explanations of quasars.

radiation pressure   the tiny pressure exerted on a surface by light or other electromagnetic radiation.

radioactivity   spontaneous disintegration of certain nuclei with the emission of *alpha*, *beta* or *gamma* radiation.

radiometer effect   toy radiometers rotate in the opposite direction to that expected from *radiation pressure*. This is a consequence of the local heating of residual gas in the device.

red-shift   the displacement of the wavelength of received light to the red end of the spectrum, compared with its emission wavelength. The most common cause of the red-shift is the *Doppler effect*, due to the source of light moving away from the receiver. *Gravity* also induces a red-shift when light is emitted from a massive body.

relativity   fundamental theory of *space-time*. *Special relativity* considers all processes excluding *gravity*, which is the province of *general relativity*.

Schwarzschild radius   the radius of the *event horizon* of a *black hole*.

solar wind   a flow of energetic particles, mostly *protons* and *electrons*, from the outer atmosphere of the Sun.

space-time   the four-dimensional space-time stage for events in *relativity* theory.

space-time diagram   pictorial representation of events in *space-time*, where time is plotted vertically and space horizontally.

special relativity

fundamental principles of physics formulated within the *space-time* frameworks of uniformly moving observers. Such a framework is only possible in the absence of *gravity*. (*General relativity* is appropriate in situations where gravity is present.) Special relativity is based on two postulates: (1) the speed of light in a vacuum is a universal constant, independent of the motion of the source or receiver; (2) the laws of physics are the same for all uniformly moving observers.

spectra

passing ordinary white light through a prism reveals that it consists of a range of different colours from red to violet – its spectrum. The whole *electromagnetic spectrum* extends this range into invisible colours of both longer and shorter wavelength. The spectra obtained from various luminous sources also often contain bright 'lines', called *emission spectra* or dark 'lines' called *absorption spectra*.

Standard Model

the reasonably well established part of particle physics theory. This corresponds to the theory of *Quantum chromodynamics* for *quark* strong interactions plus the *electroweak theory* for *electromagnetism* and the *weak force*.

string theory

the idea that the most basic entities are not point particles, but rather are objects with extension in one dimension like a string. Such theories have formed the basis of a new theory of *gravity* and the unification of all four forces of nature.

strong nuclear force/strong interaction

the force that holds the nucleus together. More generally, it is a remnant of the inter-*quark* 'colour' force and acts on all particles made of quarks.

supernova

a catastrophic stellar explosion in which the power output becomes comparable with that of a whole galaxy for about a month and thereafter gradually declines. The explosion may leave behind a *neutron star* or *black hole*.

supersymmetry

an abstract transformation which links the force particles – such as photons – to the matter particles – such as electrons. (The particles linked are not as simple as this.) Considering theories which have this transformation as a symmetry has been a basic ingredient in many attempts to provide a quantum theory of gravity, including the *string theories* called superstrings.

Theories of Everything (TOE)

theories purporting to unify all the forces and particles of nature. Superstring theory is one such attempt at the complete unification of physics.

thermodynamics

the science of the relationships between heat, work and other forms of energy. The two key laws of thermodynamics are: (1) heat is a form of energy, and energy cannot be created or destroyed (conservation of energy); (2) heat flows spontaneously from a hot body to a colder body. (This law can be formulated in terms of the concept of *entropy*, and its atomic interpretation suggests that the disorder in the universe is always increasing.)

twin paradox

special relativity predicts that clocks, including biological clocks, carried on a round trip will lose time compared with those that remain at the starting point. This is a fact and not a paradox, although it affronts common sense.

unified field theory

any field theory which relates apparently different phenomena. Einstein spent much of the later part of his life unsuccessfully trying to construct a unified field theory for *electromagnetism* and *gravity*.

vacuum tube an evacuated tube which is able to transmit an electric current.

weak nuclear force/weak interaction one of the fundamental forces, which is responsible for beta decay and any interaction involving neutrinos.

white dwarf a dense compact remnant star with a typical mass comparable with that of the Sun but a size similar to the Earth. The star is devoid of nuclear interactions and is supported by the Pauli principle applied to electrons. It is white with remnant heat, but is gradually cooling.

world line the four-dimensional *space-time* path of an object.

Quotations and sources

We have given the page number of the quotation, along with the first few words, followed, where possible, by the primary source. Many of the original quotations can be found in the books listed in the Suggestions for further reading.

Besides the sources listed below and in the figure captions, the authors and Cambridge University Press would like to make acknowledgement to any they may have failed to trace.

The Albert Einstein quotations are reprinted by permission of The Albert Einstein Archives, The Jewish National and University Library, The Hebrew University of Jerusalem, Israel.

Preface

p. xi   'In general ...', Richard Feynman, *Messenger Lecture* (The Character of Physical Law, MIT Press, 1967), p. 156.

'Ah, what a foolish ...', Johannes Kepler, *Astronomia Nova* , in Arthur Koestler, *The Sleepwalkers* (Penguin Books, Harmondsworth, 1964).

Chapter 1

p. 1   'My solution was ...', Albert Einstein, *Kyoto Address*, 1922, in T. Ogawa, *Japan St. Hist. Sci.* **18**, (1979).

'The wonderful events ...', Albert Einstein, *I. Newton, Optics* (McGraw-Hill, New York, 1931).

p. 2   'Soon I will arrive ...', Albert Einstein, *Lettres á Maurice Solovine*, trans. M. Solovine (Paris, 1956), p. 5.

p. 3   'immediately abandon ...', in Lewis Feuer, *Einstein and the Generations of Science,* 2nd edn (Transaction Books, New Brunswick, 1982), p. 45.

p. 4   'By an application ...', Albert Einstein, *The Times*, November 28th, 1919.

'We concede ...', *The Times*, November 28th, 1919.

'sometimes we "wonder" ...', Albert Einstein, *Autobiographical notes: Albert Einstein: Philosopher/Scientist*, ed. P. A. Schilpp (Tudor, New York, 1949), p. 5.

p. 6   'The normal adult ...', C. Seelig, *Einstein: A Documentary Biography* (Tr. Mervyn Savill, London, 1956), pp. 70–1.

p. 8   'I was sitting ...', Albert Einstein, *Kyoto Lecture*, in J. Ishiwara, *Einstein – Koen-Roku* (Tokyo-Tosho, 1977).

p. 9   'It might appear possible ...', Albert Einstein, *Ann. der Physik (Leipzig)* **17**, 891 (1905).

p. 17   'We conclude that ...', Albert Einstein, *ibid.*

Chapter 2

p. 23   'The factor which ...', Albert Einstein, *Autobiographical notes: Albert Einstein: Philosopher/Scientist,* ed. P. A. Schilpp (Tudor, New York, 1949), p. 25.

'That gravity should ...', Isaac Newton, 3rd letter to Richard Bentley, *c*1692–3, reprinted in I. B. Cohen, *Isaac Newton's Papers and Letters on Natural Philosophy and Related Documents* (Harvard University Press, 1958).

p. 36   'As a student …', Albert Einstein, *Kyoto Address*, 1922, in T. Ogawa, *Japan St. Hist. Sci.* **18**, 73 (1979).

         'I have been observing …', Ole Roemer, 'Proof of the movement of light', *Journal des Scavans* (1676).

p. 39   'I am not in the least …', George Francis Fitzgerald, letter to O. Heavyside, *c*1900.

p. 40   'The result of the hypothesis …', A. A. Michelson, *Am. J. Sci.* **22**, 120 (1881).

p. 42   'For more than two generations …', Robert Cecil, Marquis of Salisbury, Address to the British Association, 1894.

         'The result …', A. A. Michelson, *Am. J. Sci.* **22**, 120 (1881).

         'Perhaps we must …', H. Poincaré, Address to the International Congress of Arts and Science at St Louis, *Bull. Sci. Math.* **28**, 302 (1904).

         'Lorentz had recognized …', Albert Einstein, letter to C. Seelig, February 19th, 1955, in Carl Seelig, *Albert Einstein: A Documentary Biography* (Europa Verlag, Zurich, 1960), p. 114.

p. 43   'I have for long thought …', John Bell, *Speakable and Unspeakable in Quantum Mechanics* (Cambridge University Press, 1987), p. 67.

Chapter 3

p. 46   'The introduction of …', Albert Einstein, *Ann. der Physik (Leipzig)* **17**, 891 (1905).

p. 48   'People give ear …', Martin Luther, *c*1533, translated and quoted by Andrew D. White, *A History of the Warfare of Science with Technology in Christendom*, vol. I (Appleton, New York, 1896), p. 126.

p. 49   'Recently I have …', Albert Einstein, *Kyoto Address*, 1922, in T. Ogawa, *Japan St. Hist. Sci.* **18**, 73 (1979).

         'Thank you …', Albert Einstein, *ibid.*

p. 50   'At last …', Albert Einstein to R. S. Shankland, in R. S. Shankland, *Conversations with Albert Einstein*, (American Institute of Physics, New York, 1963), **31**, p. 47.

         'Do you know …', David Hilbert, *Einstein: His Life and Times* (Frank, New York, 1947), p. 206.

         'that he had also …', Albert Einstein to Abraham Pais, in Abraham Pais, *Subtle is the Lord …'* (Oxford University Press, 1982), p. 117.

         'was so overwhelming …', Abraham Pais, *ibid.*, p. 173.

p. 56   'Henceforth space …', Herman Minkowski, *Space and Time*, quoted in Albert Einstein *et al.*, *The Principle of Relativity* (Dover, New York, 1923).

         'Oh, that Einstein …', Herman Minkowski.

p. 57   'I do not know …', David Hilbert, quoted in Theodore von Karman, *The Wind and Beyond*, p. 51.

         'What a pity …', Herman Minkowski's alleged last words, January 1909, quoted in R. W. Clark, *Einstein: The Life and the Times* (Hodder & Stoughton, London, 1973).

         'The non-mathematician …', Albert Einstein, *Relativity* (University Paperbacks, Methuen, London, 1960), p. 55.

p. 64   'What led me …', Albert Einstein.

         'If we placed …', Albert Einstein, 1911, in G. J. Whitrow, *The Nature of Time* (Penguin, 1975), p. 93.

p. 66   'adopted Poincaré's …', Sir Edmund Whittaker, *Biog. Mem. Fellow Roy. Soc.* **1**, 37 (1955).

p. 67   'Although I was …', Max Born, *Physics in my Generation* (Publisher, London, 1956), pp. 193–5.

Chapter 4

p. 68   'After ten years …', Albert Einstein, *Autobiographical notes: Albert Einstein: Philosopher/Scientist*, ed. P. A. Schilpp (Tudor, New York, 1949), p. 53.

p. 78   'like coming upon …', William Huggins, quoted in Georgio Abetti, *The History of Astronomy*, trans. Betty Burr Abetti (Schuman, New York, 1952), p. 192.

p. 81   'He had steeled himself …', Isaac Asimov, *Foundation* (Grafton Books, 1960), p. 10.

Chapter 5

p. 88   'The most important …', Albert Einstein, *Relativity* (Methuen, London, 1960), p. 45.

p. 89   'Chemists have turned …', Antoine Laurent Lavoisier, quoted in J. Schwinger,
Einstein's Legacy – The Unity of Space and Time (Scientific American Library, Scientific American Book Inc., 1986), p. 90.

p. 91   'I was struck …', Benjamin Thomson, Count Rumford, 'Enquiry concerning the source of heat which is
excited by friction', Paper to the Royal Society, 1798.
'any thing which …', Benjamin Thomson, Count Rumford, *ibid.*

p. 92   'force, once in existence …', Julis Robert Mayer, in J. C. Foster, *Phil. Mag.* **24**, 371–7 (1862).

p. 97   'That Professor Goddard …', Editorial, *New York Times*, January 13th, 1920.

p. 98   'The measurements …', W. Kaufman, *Ann. der Physik (Leipzig)* **19**, 487 (1906).
'It is … noted …', Albert Einstein, *Jahrb. Rad. Elektr.* **4**, 411 (1907).

p. 100   'that light transfers mass …', Albert Einstein, letter to Conrad Habicht, 1905, quoted in Carl Seelig,
Albert Einstein: A Documentary Biography (Europa Verlag, Zurich, 1960), p. 353.

Chapter 6

p. 105   'It is possible …', Albert Einstein, *Jahrb. Rad. Elektr.* **4**, 411 (1907).

p. 106   'By and by …', Albert Einstein, *Autobiographical notes: Albert Einstein: Philosopher/Scientist*, ed.
P. A. Schilpp (Tudor, New York, 1949), p. 53.

p. 107   'It seems probable …', Isaac Newton, *Optics*, (McGraw-Hill, New York, 1931).
'What the atom …', Robert Cecil, Marquis of Salisbury, Address to the British Association, 1894.

p. 111   'in a purely mechanical world …', Friedrich Wilhelm Ostwald, *Vehr. Ges. Deutsch. Naturf. Arzte*. **1**, 155 (1895).
'if faith in the reality …', Ernst Mach, *Scientia*, vol. 7 (1910), p. 233.

p. 112   'entitle even the cautious …', Friedrich Wilhelm Ostwald, *Grundriss ser Physikalsiken Chemic* (Grossbothen, 1908).
'The 2nd law of thermodynamics …', James Clerk Maxwell, in Rayleigh, *Life of Lord Rayleigh*
(University of Wisconsin Press, 1968), p. 47.
'the almost unanimous …', J. J. Thomson, quoted in G. K. T. Conn and H. O. Turner, *The Evolution of the
Nuclear Atom* (Elsevier, New York, 1965), p. 53.

p. 114   'On the contrary …', Henri Becquerel, quoted in Emilio Segrè, *From X rays to Quarks* (W. H. Freeman,
San Francisco, 1980), p. 29.

p. 117   'we retired about …', Ernest Rutherford, quoted in A. S. Eve, *Rutherford* (Macmillan, New York, 1939), p. 93.

p. 118   'It is conceivable …', Pierre Curie, in *Nobel Lectures in Physics, 1901–22* (Elsevier, New York, 1967).
'the space occupied …', Philip Lenard, quoted in Richard Rhodes *The Making of the Atomic Bomb*
(Simon & Schuster, New York, 1986).
'There are at present …', Ernest Rutherford, *Phil. Mag.* **47**, 109 (1899).

p. 119   'I have never had …', J. J. Thomson, quoted in Emilio Segrè, *From X rays to Quarks* (W. H. Freeman,
San Francisco, 1980), p. 51.

p. 120   'conveyed the tremendous …', Frederick Soddy, *Atomic Transmutation* (New World, 1953), p. 126.
'The energy of …', Ernest Rutherford, *The Collected Papers, vol. I.* (Allen and Unwin, London, 1962), pp. 606ff.

p. 125   'Then a sudden …', Ernest Rutherford, quoted in J. D. Burchfield, *Lord Kelvin and the Age of the Earth* (Science History
Publication, New York, 1975).
'It was almost as if …', Ernest Rutherford, quoted in E. N. da C. Andrade, *Rutherford and the Nature of the Atom*
(Doubleday, New York, 1963), p. 111.
'whose short duration …', Niels Bohr, *Collected Works, vol. I* (North Holland, New York, 1972), pp. 98–9.
'It takes half a year …', Niels Bohr, *Oral History Interview*, Center for the History of Physics (American Institute of
Physics, New York), pp. 13–14.

p. 122   'the patience to listen …', Niels Bohr, *Essays 1958–1963 on Atomic Physics and Human Knowledge*
(Interscience, New York, 1963), p. 32.

p. 122   'Bohr's different …', Ernest Rutherford, quoted in Stefan Rozental (ed.), *Niels Bohr* (North Holland, New York, 1967), p. 46.

'It could be …', Niels Bohr, letter to his brother, 1912. This translation was constructed from the most idiomatic phrases from each translation of Niels Bohr, *Collected Works, Vol. I* (North Holland, New York, 1972), p. 559; and J. L. Heilbron and T. S. Kuhn, *The Genesis of the Bohr Atom,* Historical Studies in the Physical Sciences 1:211 (1969), p. 238.

'As soon as I saw …', J. L. Heilbron and T. S. Kuhn, *ibid.*

p. 124   'The first, for me …', Werner Heisenberg, in *From a Life in Physics, IEAE* Bulletin, 1968, p. 35.

'We are now …', Werner Heisenberg, quoting Niels Bohr, *ibid.*

p. 125   'I have already sent …', Paul Ehrenfest, in G. E. Uhlenbech, *Phys. Today* **29**, 43 (1976).

p. 126   'He is a second Dirac …', Eugene Wigner on Richard Feynman. From a letter from Robert Oppenheimer to Birge, in Smith and Weiner, *Robert Oppenheimer: Letters and Reflections* (Harvard University Press, 1980), p. 269.

'a whole series …', Paul Dirac, *Principles of Quantum Mechanics* (GTTI, 1932), publisher's preface.

p. 128   'perhaps the most …', John van Vleck, quoted by P. T. Mathews, in B. N. Kursunoglu and E. P. Wigner, *Paul Adrien Maurice Dirac* (Cambridge University Press, 1987).

'Whenever a physicist …', Wolfgang Pauli, quoted in Helge S. Kragh, *Dirac: A Scientific Biography* (Cambridge University Press, 1990), p. 101.

'The saddest chapter …', Werner Heisenberg, letter to Wolfgang Pauli, July 31st, 1928.

p. 129   'a new kind of particle …', P. A. M. Dirac, *Proc. Roy. Soc. A.* **126**, 360 (1931).

p. 131   'But no matter …', Richard Feynman, *QED: The Strange Theory of Light and Matter* (Princeton University Press, 1985), p. 128.

'we should no longer …', Paul Dirac, quoted by L. M. Brown and H. Rosenberg, in B. N. Kursunoglu and E. P. Wigner, *Paul Adrien Maurice Dirac* (Cambridge University Press, 1987), p. 149.

p. 132   'We must regard …', Paul Dirac, *Nobel Lecture* (Norstedt and Soner, 1934), p. 768.

p. 133   'It looks like …' and 'At least …', Wolfgang Pauli and Paul Dirac, quoted by A. D. Krisch, in B. N. Kursunoglu and E. P. Wigner, *Paul Adrien Maurice Dirac* (Cambridge University Press, 1987), p. 52.

'I suppose …', Paul Dirac, quoted by R. E. Peierls, in B. N. Kursunoglu and E. P. Wigner, *ibid.*, p. 43.

Chapter 7

p. 134   'I do not believe …', Albert Einstein, *Atlantic Monthly*, 1945.

p. 135   'some fool …', Ernest Rutherford, quoted in A. S. Eve, *Rutherford* (Macmillan, 1939), p. 102.

'It is probable that …', Frederick Soddy, *Atomic Transmutation* (New World, 1953), p. 95.

p. 136   'most heavily stored …', H. G. Wells, *The World Set Free*, p. 10. In series *The Works of H. G. Wells*, Atlantic Edition, vol. XXI (T. Fisher Unwin Ltd., London, 1926).

'nobody else …', H. G. Wells quoted in series *The Works of H. G. Wells, ibid.*, p. ix.

'I wish, Daddy, …', H. G. Wells' son to his father, *ibid.*, p. 17.

p. 140   'neon consisted …', Francis Aston, in Joseph Needham and Walter Pagel (eds.), *Background to Modern Science* (Macmillan, 1938), pp. 139–40.

'Thus to change …', Francis Aston, *ibid.*, p. 140.

p. 141   'I know what those particles are', and ensuing conversation. Ernest Rutherford and Mark Oliphant, quoted in Emilio Segrè, *From X rays to Quarks* (W. H. Freeman & Co., New York, 1980), p. 117.

'There are those …', Francis Aston, in Joseph Needham and Walter Pagel (eds.), *Background to Modern Science* (Macmillan, London, 1938), pp. 113–14.

'the possible existence …', Ernest Rutherford, 'The Bakerian Lecture: Nuclear constitution of atoms', Royal Society, June 1920, in E. Rutherford, *Collected Papers vol. III* (Interscience, New York, 1965), pp. 14 ff.

p. 142   'I don't believe it!', Ernest Rutherford, quoted in James Chadwick, *Some Personal Notes on the Search for the Neutron* (Hermann, 1964), p. 161.

'What fools! …', Ettore Majorana, quoted in Emilio Segrè, *From X rays to Quarks* (W. H. Freeman & Co., 1980), p. 183.

'It was a strenuous time', James Chadwick, *Oral History Interview* (American Institute of Physics, New York), p. 71.

p. 143   'Now I want …', James Chadwick, quoted in C. P. Snow, *The Physicists* (Little Brown, New York, 1981), p. 35.

           'the prehistory …', Hans Bethe, *Oral History Interview* (American Institute of Physics, New York), p. 3.

p. 144   'a number of …', Ernest Rutherford, *Collected Papers vol. II* (Interscience, New York, 1963), p. 547.

           'We must conclude …', Ernest Rutherford, *Phil. Mag.* **37**, 581 (1919).

p. 146   'Marie Curie saw …', Frederic Joliot, quoted in Pierre Biquard, *Frederic Joliot-Curie* (Paul S. Eriksson, 1962), p. 33.

p. 148   'the atomic number …', Enrico Fermi, *Collected Papers* (University of Chicago Press, 1962), p. 750).

           'As a consequence …', O. Hahn and F. Strassman, *Naturwiss.* **27**, 11 (1939); a translation by Hans G. Graetzer may
be found in *Am. J. Phys.* **32**, 10 (1964).

           'Perhaps you can suggest …', Otto Hahn, letter to Lise Meitner, 1938, quoted in Otto Hahn *Erlebnisse und Erkenntnisse*
(Econ Verlag, 1975), p. 81.

           'We cannot hush up …', Otto Hahn, letter to Lise Meitner, 1938, quoted in Otto Hahn, *ibid.*

p. 149   'I had hardly begun …', Otto Frisch, quoted in Stefan Rozental (ed.), *Niels Bohr* (North Holland, Amsterdam, 1967), p. 145.

           'It is reasonable …', Enrico Fermi, 1934, *Collected Papers* (University of Chicago Press, 1962), p. 734.

p. 150   'A chain reaction …', Leo Szilard, 1934 patent, quoted in Leo Szilard, *The Collected Works: Scientific Papers*
(MIT Press, Cambridge, MA, 1972), p. 639.

           'In an enterprise …', Emilio Segrè, quoted in Richard Rhodes, *The Making of the Atom Bomb*
(Simon & Schuster, New York, 1986).

           'might never have …', Hans Bethe, *Oral History Interview* (American Institute of Physics, New York), p. 30.

p. 151   'No, I do not want …', Enrico Fermi, quoted in Emilio Segrè, *Enrico Fermi, Physicist*
(University of Chicago Press, 1970), p. 80.

p. 152   'It was the elimination …', Kurt Diebner, quoted in David Irving, *The Virus House* (William Kimber, 1967);
in the USA, this book has the title *The German Atomic Bomb*, and is published by Simon & Schuster.

p. 153   'was wild enough …', Louis Turner, quoted in Spencer R. Weart and Gertrud Weiss Szilard (eds.),
Leo Szilard: His Version of the Facts (MIT Press, Cambridge, MA, 1978), p. 126.

p. 154   'neutron bombardment …', Carl von Weizsacker (paper dated July 17th, 1940), quoted in T. Powers, *Heisenberg's War*
(Jonathan Cape, London, 1993), p. 83.

           'undergoes fission …', Glen T. Seaborg, *Early History of Heavy Isotope Production at Berkeley*
(Lawrence Berkeley Laboratory, 1976), p. 34.

           'Suppose someone …', Otto Frisch and Rudolf Peierls, *ibid.*

           'We estimated …', *ibid.*

           'on the back of the proverbial envelope', Rudolph Peierls, *ibid.*

p. 155   'one might produce …', Otto Frisch, *Oral History Interview* (American Institute of Physics, New York), p. 39.

           'Even if this plant …', Otto Frisch and Rudolf Peierls, quoted in Rudolf Peierls, *Bird of Passage*
(Princeton University Press, 1985), p. 323.

           'the centre of a big city', Frisch-Peierls memorandum, 1940, quoted by Margaret Gowing, in R. H. Dalitz and
R. B. Stinchcombe (eds.), *A Breadth of Physics* (World Scientific, Singapore, 1988), p. 145.

p. 156   'Since the separation …', *ibid.*, p. 147.

           'enriched uranium [was] …', Werner Heisenberg, report to the German War Office, December 1939, quoted in
David Irving, *The Virus House* (William Kimber, 1967); in the USA this book has the title *The German Atomic Bomb* and is
published by Simon & Schuster.

           'those who choose …', Rudolf Peierls (personal communication to A. J. G. Hey, 1988): see also R. E. Peierls and N. Mott,
Werner Heisenberg, *Biog. Mem. Fellow Roy. Soc.* **23**, 213 (1977).

p. 157   'promised to get it to the right person', Mark Oliphant, quoted in Rudolf Peierls, *Bird of Passage*
(Princeton University Press, 1985), p. 155.

           'one of the most …', Margaret Gowing, in R. H. Dalitz and R. B. Stinchcombe (eds), *A Breadth of Physics*
(World Scientific, Singapore, 1988), p. 148.

p. 157   'PLEASE INFORM …', Niels Bohr, telegram, quoted in Ronald Clark, *The Greatest Power on Earth* (Harper & Row, 1980), p. 95.

Chapter 8

p. 161   'In 1907 …', Albert Einstein, 1922, quoted in J. Ishiwara, *Einstein – Koen-Roku* (Tokyo-Tosho, 1977).

'What do colleagues …', Albert Einstein, letter to L. Hopf, August 16th, 1912.

'As an older friend …', Max Planck to Albert Einstein, quoted in a letter from E. G. Strauss to A. Pais, October 1979.

p. 163   'If you were standing …', Richard Feynman, *The Feynman Lectures in Physics*, vol. II (Addison-Wesley, Reading, Mass., 1964), p. 1.

p. 165   'it was clear …', Max Abraham, *Ann. der Physik*. **38**, 1056 (1912).

'a stately horse …', Albert Einstein, letter to L. Hopf, August 16th, 1912.

p. 167   'the happiest thoughts of my life …', Albert Einstein, *Morgan Manuscript* (Pierpont Morgan Library, New York, 1920).

'The gravitational field …', Albert Einstein, *ibid.*

p. 170   'Because for an observer …', Albert Einstein, *ibid.*

Chapter 9

p. 181   'No one who has …', Albert Einstein, quoted by A. Pais, *Subtle is the Lord* (Oxford University Press, 1982).

'Help me, Marcel, or I'll go crazy!, quoted in P. Michelmore, *Einstein: Profile of the Man* (Dodd, Mead, New York, 1962), p. 60.

'I have made …', Janos Bolyai, letter to his father, November 3rd, 1823.

p. 182   'One geometry …', Henri Poincaré, *Science and Hypothesis* (Dover, New York, 1952), p. 50.

p. 183   'If all [accelerated] systems …', Albert Einstein, *Kyoto Lecture*, 1922, in J. Ishiwara, *Einstein – Koen-Roku* (Tokyo-Tosho, 1977).

'I am now exclusively occupied …', Albert Einstein, letter to A. Sommerfeld, October 29th, 1912.

p. 185   'I became more and more convinced …', Karl Friedrich Gauss, letter, 1817, in *Werke* (Konigliche Gesellschaft der Wissenschaften, Göttingen, 1900).

p. 186   'Indeed I have from time to time …', Karl Friedrich Gauss, letter, 1824, *ibid.*

'the clamour of the Boeotians', Karl Friedrich Gauss, *ibid.*

'For God's sake …', Farkas Bolyai, letter to his son, Janos, 1825

p. 188   'the happy achievement …', Albert Einstein, *Notes on the Origin of the General Theory of Relativity* (Jackson, Wylie, Glasgow, 1934).

p. 190   'To my great joy …', Albert Einstein, letter to W. J. de Haas, undated but probably August 1915.

'Every boy in the streets of Göttingen …', David Hilbert, quoted in Constance Reid, *Hilbert* (Allen and Unwin, London, 1970).

p. 191   'Dear Mother, …', Albert Einstein, postcard to his mother, September 27th, 1919.

p. 194   'a tremendous rainstorm came on', Arthur Eddington, quoted in A. V. Douglas, *The Life of A. S. Eddington* (Thomas Nelson, London, 1956).

'That was the happiest moment of my life!', Arthur Eddington, *ibid.*

'The evidence …', Arthur Eddington, *ibid.*

'We owe it to that great man …', Ludwick Silberstein, *Proc. Roy. Soc.* **96A**, p. i.

'Lights all askew …', *New York Times*, November 9th, 1919.

'My aim lies in smoking …', Albert Einstein, *Kyoto Lecture*, 1922, in J. Ishiwara, *Einstein – Koen-Roku* (Tokyo-Tosho, 1978).

p. 199   'The final release …', Albert Einstein, letter, 1915.

'if I had believed …', Johannes Kepler, *Astronomia Nova*, in Arthur Koestler, *The Sleepwalkers* (Penguin Books, Harmondsworth, 1964).

p. 200   'I thought and searched …', Johannes Kepler, *ibid.*

'there is a force …', *ibid*, p. 338.

p. 202   'without any special hypothesis', Albert Einstein, *Silzungberichte,* Preussische Akademic der
           Wissenshaften, 1915, p. 831.
           'This discovery was …', Abraham Pais, *Subtle is the Lord* (Oxford University Press, 1982).

Chapter 10
p. 210   'The sceptic will say …', Albert Einstein, *Sci. Am.,* April, 1950, p. 17.
           'While on Mount Wilson …', R. Berendzen, R. Hart and D. Seeley, *Man Discovers the Galaxies,*
           Science History Publications (Neale Watson Academic Publishers, New York, 1976), p. 200.
           'I have … again …', Albert Einstein, letter to Paul Ehrenfest, February 4th, 1917.
p. 212   'The results obtained …', Albert Einstein, *Zeits. Physik.* (1922).
           'We have Einstein's space …', Sir J. J. Thomson, *Recollections and Reflections* (Bell & Sons, London, 1939), p. 431.
p. 214   'which the general theory …', Albert Einstein, *Sitzungberichte,* Preussische Akademic der Wissenshaften, 1931, p. 235.
p. 216   'One puzzling question …', Arthur Eddington, quoted in *The Cosmos (Voyage through the Universe)*
           (Time Life Books, Alexandra, Virginia, 1989), p. 62.
p. 217   'ashes and smoke …', Georges Lemaître, quoted in *The Cosmos …, ibid.*
p. 219   'Thus, you see …', George Gamow, letter to Arno Penzias, September 28th, 1965. See Fig. 10.12.
p. 220   'the loss of the Steady State theory …', Dennis Sciama, in R. Kippenhahn, *Light from the Depths of Time*
           (Springer Verlag, Berlin, 1967), p. 153.
p. 223   'The star thus tends …', J. Robert Oppenheimer and Hartland Snyder, *Phys. Rev.* **54**, 455 (1939).
p. 224   'The consequences of my crime …', John Wheeler, *A Journey into Gravity and Spacetime* (Scientific American Library,
           New York, 1990), p. 221.
           'You don't destroy entropy …', Jacob Berkenstein, in John Wheeler, *ibid.*
p. 230   'the first known system …', Clifford Will, in S. W. Hawking and W. Israel (eds.), *General Relativity, an Einstein
           Centenary Survey* (Cambridge University Press, 1979), chap. 2.
p. 233   'I don't like …', Richard Feynman, in P. C. W. Davies and J. Brown, eds., *Superstrings: A Theory of Everything?*
           (Cambridge University Press, 1988), p. 194.
p. 234   'I have become …', Albert Einstein, letter to a friend, 1942.
           'In regard to work …', Albert Einstein, *ibid.*
p. 235   'a diamond is forever', Sheldon Glashow, seminar (date unknown).
           'In the judgement …', Albert Einstein, letter to the *New York Times,* May 4th, 1935.
           'The idea of achieving …', Albert Einstein, letter to Theodor Kaluza, April 21st, 1919.
p. 236   'I remember that …', Abraham Pais, *Subtle is the Lord* (Oxford University Press, 1982), footnote, p. 332.
           'I must seem …', Albert Einstein, letter to Louis de Broglie, February 8th, 1954.
           'I have had my first good idea in ten years!', Peter Higgs, note to Lewis Ryder, (personal communication to, A. J. G. Hey,
           1993).
p. 238   'Einstein developed …', Edward Witten, in P. C. W. Davis and J. Brown, eds., *Superstrings: A Theory of Everything?*
           (Cambridge University Press, 1988), p. 102.
           'I'm particularly annoyed …', Sheldon Glashow, in P. C. W. Davis and J. Brown, *ibid.*
p. 239   'only the germ of an idea', Roger Penrose, *The Emperor's New Mind* (Oxford University Press, 1989).
           'he stopped thinking …', Richard Feynman, quoted by Freeman Dyson, *Disturbing the Universe*
           (Harper & Row, New York, 1979), p. 62.
           'Above stands …', Albert Einstein, letter to H. Weyl, May 26th, 1923.

Chapter 11

p. 240   'Extravagant Fiction …', *Amazing Stories*, April 1926, title page.

'divorced astronomy …', Arthur Koestler, *The Sleepwalkers* (Penguin Books, Harmondsworth, 1964), p. 318.

p. 241   'The initial shock …', Johannes Kepler, *Somnium*, ed. and trans. by Ludwig Gunther (Leipzig, 1818).

p. 242   'an argument …', Johannes Kepler, *ibid.*, footnotes.

'I have just finished …', Jules Verne, 1863.

'to ascertain the shock …', Jules Verne, *From the Earth to the Moon* (Digit, London, 1958).

'At the moment …', Jules Verne, *ibid.*, p. 151.

p. 243   'There are really …', H. G. Wells, *The Time Machine* (1895), in *Selected Short Stories of H. G. Wells* (Penguin, Harmondsworth, 1958), p. 8.

p. 244   'you can move about …', H. G. Wells, *ibid.*, p. 10.

'That is the germ …', H. G. Wells, *ibid.*

'He pointed …', H. G. Wells, *The First Men in the Moon* (1901) (Everyman, London, 1943), p. 33.

'flying-machine that will …', H. G. Wells, *The Argonauts of the Air*, *Phil. Mag. Annual,* 1895; also in *Selected Short Stories of H. G. Wells* (Penguin, Harmondsworth, 1958).

p. 245   'A good science fiction story …', Frederick Pohl, quoted in Larry Niven and Jerry Pournelle, *The Mote in God's Eye* (Futura Publications, London, 1975).

p. 246   'Throughout the past …', Larry Niven and Jerry Pournelle, *ibid.*, prologue.

p. 247   'It takes …', Larry Niven and Jerry Pournelle, *ibid.* p. 32.

p. 248   'Thanks to its perfected rockets …', Pierre Boulle, *La Planete des Singes* (1963); translated as *Monkey Planet* (Penguin, Harmondsworth, 1966), p. 12.

p. 250   'He found that there …', Larry Niven, *Flash Crowd* (1973), in *The Plight of the Horse* (Futura, London, 1975), p. 113.

p. 251   'Given; a burnt out white dwarf …', Larry Niven, *Neutron Star* (1966), (Futura, London, 1969), p. 14.

p. 252   'Well, we've got …', Gregory Benford, *Timescape* (1980) (Bantam, New York, 1992), p. 10.

p. 253   'Professor Thorne believes …', Kip Thorne, press release, 1988; quoted in Kip Thorne, *Black Holes and Time Warps* (W. W. Norton & Co. Inc., New York, 1994), p. 516.

p. 255   'After I gave him …', Robert Forward, *Timemaster* (Tom Doherty Associates, Inc., New York, 1992).

p. 256   'With the warpgate …', Robert Forward, *ibid.*

'Once any time machine …', Robert Forward, *ibid.*

'If time machines …', Kip Thorne, *Black Holes and Time Warps* (W. W. Norton & Co. Inc., New York, 1994), p. 516.

Figures

Figure 1.2   'when we ran …', Albert Einstein, letter to Maurice Solovine, November 25th, 1948.

Figure 1.8   'I could not have got …', Albert Einstein, letter to Michel Angelo Besso, 1919.

'that cloister …', Albert Einstein, *ibid.*

Figure 2.1   'Nature and Nature's laws …', Alexander Pope, 1727.

Figure 2.4   'What good is a newborn baby?', Michael Faraday, *c* 1840, quoted in William J. Kaufmann III, *Galaxies and Quasars* (W. H. Freeman & Co., San Francisco, 1979), p. 208.

Figure 2.13   'The enormous significance …', Albert Einstein, *Math. Naturw. Blatt.* **22**, 24 (1928).

Figure 3.4   'more scandalous …', declaration from the Catholic Church, 1633, quoted in Morris Kline, *Mathematics and Western Culture* (Pelican, London, 1972), p. 148.

Figure 5.1   'The Republic …', judges commenting on the appeal on behalf of Antoine Lavoisier, 1794, quoted in D. L. Hurd and J. J. Kipling (eds), *The Origins and Growth of Physics Science*, vol. 1 (Penguin, London, 1964), p. 324.

Figure 5.3   'Mr. Secretary-Colonel-Admiral-Philosopher Thomson', Edward Gibbon, quoted in J. Schwinger, *Einstein's Legacy – The Unity of Space and Time* (Scientific American Library, New York, 1986).

Figure 6.11 'The Professor …', Peter Kapitza, letter to his mother.

Figure 7.1 'The fireball expanded …', Leona Marshall Libby, *The Uranium People* (Crane Russack, 1979), p. 303.
'This thing is the plague of Thebes', J. Robert Oppenheimer, quoted in Richard Rhodes, *The Making of the Atomic Bomb* (Simon & Schuster, New York, 1986), p. 777.

Figure 7.17 'I opened the safes …', Richard Feynman, *Surely You're Joking Mr. Feynman* (W. W. Norton & Co. Inc., New York, 1985), p. 149.

Figure 7.24 'The first megaton-yield …', *Los Alamos Science*, 4:7, Winter/Spring 1983, p. 112.

Figure 8.2 'He snatched lightning …', Turgot, *c* 1790.

Figure 8.4 'You have written …', Emperor Napoleon, *c* 1805, quoted in A. L. Mackay, *Harvest of a Quiet Eye* (Institute of Physics, Bristol, 1977), p. 92.
'Sire, I have no need …', Marquis Pierre Simon de Laplace, *c* 1805, quoted in A. L. Mackay, *ibid.*

Figure 9.3 'The need to express …', Albert Einstein, in C. Seelig (ed.), *Helle, Zeit, dunkele Zeit* (Europa Verlag, 1956).

Figure 9.8 '[The professor said gravely] …', George Gamow, *Mr Tompkins in Paperback* (Cambridge University Press, Canto edition, 1995), p. 26.

Figure 9.9 'But … physicists …', Albert Einstein, *Essays in Science* (Philosophical Library, New York, 1934), p. 68.

Figure 9.10 'Physics is much too hard for physicists', David Hilbert, quoted in Constance Reid, *Hilbert* (Allen & Unwin, London, 1970).
'We must know. We will know.', epithet from David Hilbert's tomb, quoted in Constance Reid, *ibid.*

Figure 9.15 'A new great figure …', *Berliner Illustrirte Zeitung*, December 14th, 1919, cover.

Figure 10.30 'so magical …', Michael Green, in P. C. W. Davies and J. Brown, eds., *Superstrings: A Theory of Everything* (Cambridge University Press, 1988).

Suggestions for further reading

There are many books about Einstein and relativity. To aid the interested reader in further exploration, we list some that we have found particularly helpful.

Books about Einstein

Jeremy Bernstein, *Einstein* (Fontana, New York, 1973).

R. W. Clark, *Einstein: The Life and the Times* (Hodder & Stoughton, New York, 1973).

Lewis S. Feuer, *Einstein and the Generations of Science*, 2nd edn (Transaction Books, New Brunswick, 1982).

Abraham Pais, *Subtle is the Lord. The Science and the Life of Albert Einstein* (Oxford University Press, 1982).

Books about relativity

A. P. French, *Special Relativity* (Thomas Nelson and Sons Ltd, Walton-on-Thames, 1968).

J. Schwinger, *Einstein's Legacy – The Unity of Space and Time* (Scientific American Library, New York, 1986).

Kip S. Thorne, *Black Holes and Time Warps: Einstein's Outrageous Legacy* (W. W. Norton & Co. Inc., New York, 1994).

Clifford Will, *Was Einstein Right?* (Oxford University Press, 1988).

Relativistic fantasies

George Gamow, *Mr Tompkins in Paperback*
(Cambridge University Press, 1965).

Russell Stannard, *The Time and Space of Uncle Albert*
(Faber and Faber, London, 1989).

Russell Stannard, *Black Holes and Uncle Albert*
(Faber and Faber, London, 1991).

Other interesting source books

P. C. W. Davies and J. Brown (eds.), *Superstrings: A Theory of Everything?* (Cambridge University Press, 1988).

Arthur Koestler, *The Sleepwalkers*
(Penguin Books, Harmondsworth, 1964).

Robert Lambourne, Michael Sallis and Michael Shortland,
Close Encounters? Science and Science Fiction
(Adam Hilger, Bristol, 1990).

David Mermin, *Boojums All The Way Through*
(Cambridge University Press, 1990).

Roger Penrose, *The Emperor's New Mind*
(Oxford University Press, 1989).

Richard Rhodes, *The Making of the Atomic Bomb*
(Simon & Schuster, New York, 1986).

Richard Rhodes, *Dark Sun. The Making of the Hydrogen Bomb*
(Simon & Schuster, New York, 1995).

C. P. Snow, *The Physicists* (Little Brown, New York, 1981).

Name index

Italicized page numbers refer to entries in the Chronology section.

Subject index

Italicized page numbers refer to entries in the Chronology section; **bold** page numbers refer to Glossary entries.